AF324125

Vertical Density Representation and Its Applications

Vertical Density Representation and Its Applications

Marvin D Troutt

Kent State University, USA

W K Pang
S H Hou

The Hong Kong Polytechnic University, China

World Scientific

NEW JERSEY · LONDON · SINGAPORE · SHANGHAI · HONG KONG · TAIPEI · BANGALORE

Published by

World Scientific Publishing Co. Pte. Ltd.

5 Toh Tuck Link, Singapore 596224

USA office: Suite 202, 1060 Main Street, River Edge, NJ 07661

UK office: 57 Shelton Street, Covent Garden, London WC2H 9HE

British Library Cataloguing-in-Publication Data
A catalogue record for this book is available from the British Library.

VERTICAL DENSITY REPRESENTATION AND ITS APPLICATIONS

ISBN 981-238-693-9

Printed in Singapore by World Scientific Printers (S) Pte Ltd

Dedications

To the memory of my Mother, to my Father, to Mark, Marleen and Andrew, and to Helen.

To my parents, to my wife, Sui-Ching, to Wai-Ying and Chun-Yan.

To my parents and my family.

MDT, WKP, SHH

Preface

Since the first paper related to Vertical Density Representation (VDR) appeared in 1991 several papers have been published and work continues on the topic. VDR arose in connection with the analysis of performance measurement. Consideration of the Box-Muller method for generation of normal variates led to a question about the density (probability density function, or pdf) of the density function itself. The purpose of this book is to survey these results and provide some new unpublished results. We hope to have made contact with the major published articles on the topic. While it is possible that some could have been missed, we have made a diligent search and have sought the advice of other researchers on the topic.

VDR may be regarded as a special kind of variable transformation but may also be considered as a more general density-modeling tool. By assuming that a variate is uniformly distributed on the contours or level curves of a given function in real n-dimensional space, and considering the density of the ordinate of the given function, the density of the original variate can be represented. Basic results and extensions are discussed. Several applications are described for use with Monte Carlo simulation. The uniformly distributed on contours assumption can be relaxed with what we call the General VDR Theorem. An application of that result is made to a problem in inverse linear programming.

A new result on the analysis of correlation into two distinct components is given along with its potential value in the aggregation of experts. Also a new result on densities of orbit values produced by chaos generators is given along with the construction of a large class of chaos-based uniform random number generators. We also provide a case study in applying VDR in connection with what may be called behavioral estimation. VDR considerations are brought to bear in developing a validation technique for the estimation method. Then a further application to estimating benchmark costs and cost matrices is given. Finally in the last chapter, we discuss some future research questions and work in progress for several of the topics discussed in the earlier chapters.

In order to have a good understanding of the book, a background in mathematics and statistics at the graduate level will be best. However, readers having had a calculus-based statistics course and some acquaintance with Lebesgue measure should fare reasonably well. In the past, statistical theory has depended primarily on the Riemann integral. VDR depends more heavily on Lebesgue measure. However, the derivations are kept intuitive with respect to those issues in so far as possible.

We wish to thank Elsevier Science (http://www.elsevier.com), Taylor and Francis (http://www.tandf.co.uk) and the Decision Sciences Institute for granting us permissions to use some of the materials from the papers published in the *European Journal of Operational Research, Statistics* and *Decision Sciences* (http://decisionsciences.org) Journal. This monograph is also supported by the research grant committee of the Hong Kong Polytechnic University (Grant code: A-PD05).

We also wish to make a special acknowledgment of Professor Samuel Kotz. Without his early support and contributions to the topic, this book may not have been possible.

Marvin D. Troutt
Graduate School of Management
Kent State University
Kent, Ohio, USA

Wan-Kai Pang and Shui-Hung Hou
Department of Applied Mathematics
The Hong Kong Polytechnic University
Hong Kong, SAR,
China

August 31, 2003

Contents

Chapter 9   Open Questions and Future Research   **213**

Chapter 1

Vertical Density Representation

Introduction

The term "vertical density representation" (VDR) was first used in Troutt (1993) to describe a technique for representing densities (probability density functions or pdfs) on $\Re^n$ in terms of the ordinate of nonnegative functions on $\Re^n$. Such functions may themselves be pdfs on $\Re^n$. Although the term, VDR, was not used at that time, the first paper (Troutt, 1991) directly related to the topic focused on the latter case where a formula for the density of the density function value was derived. Namely, let $f(\mathbf{x})$ be the pdf of a random variable $\mathbf{X}$ on $\Re^n$, and consider the ordinate of the density as a random variable $V = f(\mathbf{x})$. Let $A(v)$ be the Lebesgue measure of the set $\{\mathbf{x} : f(\mathbf{x}) \geq v\}$. It was shown in that paper that if V possesses a density $g(v)$ and $A(v)$ is differentiable, then $g(v) = -vA'(v)$ on the range of $f(\mathbf{x})$. This result gives, *inter alia*, a new derivation of the Box-Muller method as will be developed below.

In Troutt (1993), a generalization was given for the case in which $V(\mathbf{x})$, a function on $\Re^n$ but not necessarily a pdf, and $g(v)$, the density of the ordinate of $V(\mathbf{x})$, are specified. It is now required to find the resulting pdf, $f(\mathbf{x})$, of $\mathbf{X}$. When that density exists and $A(v)$ is the Lebesgue measure of the set $\{\mathbf{x} : V(\mathbf{x}) \geq v\}$, it is given by $f(\mathbf{x}) = \phi(V(\mathbf{x}))$ where $\phi(v) = -g(v)/A'(v)$. The pdf $g(v)$ can be called the vertical or ordinate density. In later applications, v often represents a performance score, so that $g(v)$ may be called the performance or performance score density. There are cases when $f(\mathbf{x})$ and $V(\mathbf{x})$ are given, and it is desired to find $g(v)$. Then the problem is essentially a special kind of change of variables technique. For

1

many applications, it is the reverse situation that is of more interest. That is, we often wish to construct pdfs for special circumstances by starting with $V(\mathbf{x})$ and $g(v)$.

VDR techniques often provide a useful alternative strategy for generating random variables. Thus, most of the initial applications have been to Monte Carlo simulation. As a simple example, consider the Laplace distribution (double exponential density), $f_x(x) = \frac{1}{2}\exp(-|x|)$. The VDR or vertical density is the density of $V(x) = f_x(x)$. Here $A(v) = -2\ln(2v)$. Therefore by the formula, $g(v) = -vA'(v)$, we see that $g(v) = 2$ for $0 < v < \frac{1}{2}$. That is, $g(v)$ is the uniform distribution. Monte Carlo sampling from this density can therefore be performed as follows. First, let a value of v be sampled from the uniform density, $g(v) = 2$ for $0 < v < \frac{1}{2}$. Then either of the two corresponding x values, $\pm\ln(2v)$, can be selected with equal probability.

Kotz and Troutt (1996) applied VDR techniques to characterize the tail behavior of thirteen common univariate densities. Specifically, these were the uniform, power function, exponential, Pareto, Pearson Types II, III, and VII, normal, Cauchy, logistic, triangular, inverted triangular, and a new class proposed by N. L. Johnson. The rate of tail decrease for these densities can be ordered according to the rate of increase of the associated vertical density. That is, heaviness of the tails of a density corresponds to the degree of steepness of the vertical density for values of v near zero. For example, the Cauchy pdf has vertical density given by $g(v) = \dfrac{1}{\sqrt{\pi v(1-\pi v)}}$ for $0 < v < \pi^{-1}$. This density is unbounded as v tends to either zero or unity, corresponding to its thick-tailed behavior in the zero case, as well as, its flatness near its mode in the unity case.

The results in the early work depended on a certain uniform conditional distribution assumption. Namely, as will be discussed in detail later, it was necessary to assume that the conditional distribution of X, given $V(\mathbf{X}) = v$, which we also call the *contour density* or *pdf*, is uniform. However, in Troutt and Pang (1997), a modified VDR-type representation was obtained for the standard normal density by identifying an appropriate density on the whole of the set $A(v)$ rather than just on its boundary. Certain multivariate extensions for various L_p-norm symmetric distributions have been studied by Kotz, Fang, and Liang (1997). The uniform conditional distribution assumption is also relaxed leading to what we call the general VDR theorem.

1.1 Original Motivation

The original motivation for the work leading to VDR arose in connection
with group decision-making. Consider a group decision problem that involves choosing a most desirable vector of numbers.

Table 1.1. Relative importance of teaching, research and service

I	T	R	S
1	0.173	0.772	0.055
2	0.333	0.097	0.570
3	0.243	0.669	0.088
4	0.715	0.219	0.067
5	0.114	0.814	0.072
6	0.481	0.114	0.405
7	0.454	0.454	0.090
8	0.674	0.226	0.101
9	0.240	0.701	0.059
10	0.498	0.367	0.135
11	0.474	0.474	0.053
12	0.309	0.672	0.049
13	0.219	0.715	0.067
14	0.444	0.444	0.111
Means:	0.384	0.481	0.137

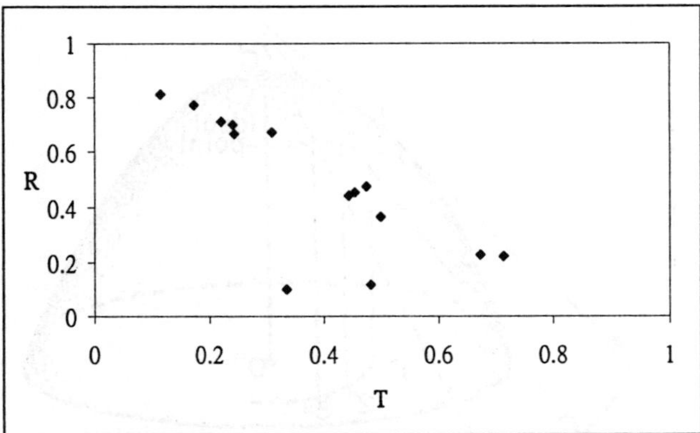

Figure 1.1: Plot of (T, R, S) data points

Table 1.1 arose in an exercise on applying the *Analytic Hierarchy Process* (Saaty, 1980). The exercise was conducted to estimate the best relative priorities on research (R), teaching (T) and service (S), respectively, for use in an academic department. Relative priorities are measures of relative importance that have been normalized to sum to unity. One might postulate the existence of a group multi-attribute value (MAV) function, $V(T, R, S) = V(\mathbf{x})$, for this decision, as suggested by Fig. 1.2. However, this MAV function is not explicitly known. Moreover, it would be quite difficult to model such a function as it would be necessary to consider all the different ways that the priority vector, (T,R,S), might be used, along with their own relative frequencies and relative importance measures. For example, such data might be used in promotion and tenure decisions, hiring decisions, merit pay recommendations, etc. However, how should those activities be weighted according to their frequencies of occurrences and their relative impacts on the welfare of the department? Given these difficulties with a direct approach to modeling an appropriate MAV function, a more expedient alternative becomes attractive. By assuming the existence of a $V(\mathbf{x})$-function, the problem becomes one of estimating its maximizer $\mathbf{x}^*$. The responses in the sample of departmental members may be thought of as individual estimates of $\mathbf{x}^*$. From that perspective, the problem becomes one of how to aggregate the individual estimates to obtain an estimate of $\mathbf{x}^*$. The problem is complicated further in that academic departments tend to be self-selecting and the members are therefore more likely to share common biases (See Figure 1.2).

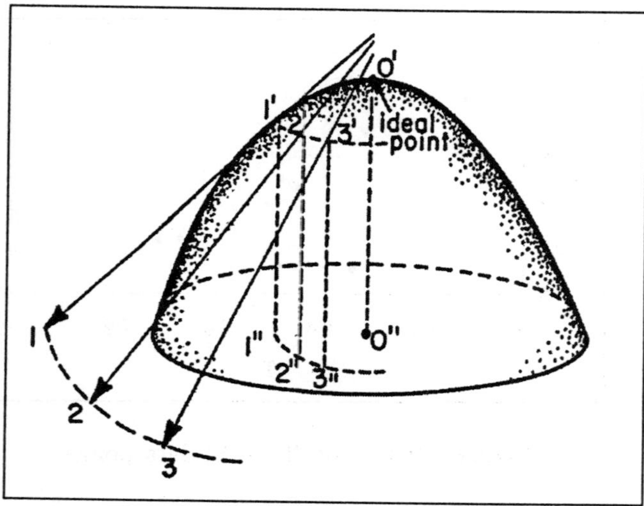

Figure 1.2: Figure for group estimation of the ideal point

Because of the potential shared bias, the usual approach of using the centroid as the aggregate estimate becomes of dubious value. That is, one would expect the centroid to reflect a similar degree of bias as that shared by individuals in the group, if any. Thus, ideally, an aggregator is desired that filters bias, or at least, suggests its presence. An early approach to the problem was proposed in Troutt (1988), which introduced the dome bias model. That model proposed one mechanism for explaining a shared group bias. The problem was further reexamined in Troutt, Pang and Hou (1999) from the point of view of mode estimation and three new such aggregators were compared for the dome bias model.

If the $V(\mathbf{x})$-function were known explicitly, then each individual estimate could be directly scored. Considering the distribution of such v-score values led to the concept of the vertical or ordinate density or pdf, $g(v)$. The question arises as to how $g(v)$ and $V(\mathbf{x})$ should be related to what may be called the *spatial* pdf $f(\mathbf{x})$ of the individual estimates. It was soon noticed that a typical unimodal $f(\mathbf{x})$ such as the multivariate normal pdf might itself serve as a $V(\mathbf{x})$-function. For that case, the question of finding $g(v)$ can be described as that of finding the density or pdf for the density or pdf itself. Together with a coincidental review of the Box-Muller method in simulation, these ideas led to the paper, Troutt (1991). After that, the more general case in which $V(\mathbf{x})$ is not necessarily a pdf was considered further in Troutt (1993), where as noted above, the VDR term was first used.

At about the same time that these ideas were developed, a related set of estimation techniques began to be considered. A first version, called maximum decisional efficiency (MDE) estimation, was proposed in Troutt (1995). A variation called maximum performance efficiency (MPE) was applied in Troutt *et al.* (2000) and Troutt *et al.* (2003). This approach to estimation focused on the achievement of the maximum average *v-score* and has the advantage that the form of $g(v)$, or $f(\mathbf{x})$, does not have to be specified in advance as in the maximum likelihood formulations. The approach also has an intuitive rationale. Namely, assuming that an appropriate model, $V(\mathbf{x})$, has been specified for the desirability of the decisions or performance measures in question, then the decision-maker or organization should have attempted to maximize its average v-score over past occasions. The approach has been applied in Troutt (1995), Troutt *et al.* (1997a, 1997b) and in Troutt *et al.* (1999). This approach is also related to frontier regression models as discussed in Troutt *et al.* (2003). The performance efficiency score, or v-score, has also been applied as a statistic to facilitate further analysis in Alsalem *et al.* (1997).

The organization of the book is as follows. Chapter 1 discusses basic results and covers the original results in the papers of Troutt (1991,1993) along with some extensions to what is called general VDR. Chapter 2 covers the results of Kotz and Troutt (1996) on applications of VDR to the ordering of distributions. This chapter also includes some new material on the analysis of correlation into two components called *vertical* and *contour* correlations. Chapter 3 reviews the results of Pang *et al.* (2001) and Kotz *et al.* (1997). This chapter deals with multivariate VDR issues. A result by Kozubowski (2002), which proved a conjecture in Troutt (1991), is also discussed. Chapter 4 is devoted to simulation applications. The results of Pang *et al.* (2002) and Fang *et al.* (2001) are described here. Chapter 5 is devoted to some new unpublished applications of VDR. It contains an application of VDR to finding pdfs of chaotic orbit values for chaos generators on the unit interval. The results are used to construct a very large class of distinct uniform random number generators based on chaos generators of a particular design.

Chapter 6 discusses some miscellaneous applications of VDR. First, we consider what we call *Tolstoy's Law*. This operationalizes a famous quote of Tolstoy (1828-1910) in his celebrated novel *Anna Karenina*. We relate this to the issue of determining when the consensus of estimators is associated with greater accuracy of the estimate. In addition, we consider a different way to generalize the normal pdf on the unit interval and compare it with the entropy approach. Next, we consider some aspects of VDR on the half-interval and connections with unimodality and Khinchin's Unimodality Theorem. These results lead to a more general result that we call Khinchin Density Representation (KDR). Finally, we discuss an application of VDR to density construction that arises in what may be called *inverse linear programming*. Chapter 7 contains a case application of VDR in what we call behavioral estimation. The *minimum decisional regret* (MDR) estimation method is discussed as an application to estimating costs related to production planning. VDR is used to construct a validation technique for that method. Chapter 8 is about the estimation of benchmark costs. Here, we define what may be called benchmark or most efficient costs per unit of cost driver. The principle of *maximum performance efficiency* (MPE) is proposed and an approach to estimating *benchmark unit costs* and *benchmark cost matrices* is derived from this principle.

1.2   The Density of the Density Ordinate

Troutt (1991) gave an alternative interpretation for the Box and Muller (1958) method for generation of normally distributed pseudo-random deviates. Recall the Box-Muller method generates normal uncorrelated deviates x_1 and x_2 from a pair of uniform $[0, 1]$ observations u_1 and u_2. Let

$$f(\mathbf{x}) = \frac{1}{2\pi} \exp\left\{ -\frac{1}{2}(x_1^2 + x_2^2) \right\}, \qquad \mathbf{x} = (x_1, x_2) \tag{1.1}$$

be the standard uncorrelated bivariate normal distribution. The method operates as follows:

(1) $h = f(\mathbf{x}) = \frac{1}{2\pi} \exp\left\{ -\frac{1}{2}(x_1^2 + x_2^2) \right\}$ may be considered as a real random variable on the real plane with range of values $\left[0, \frac{1}{2\pi}\right]$.

(2) Generate a random deviate $\frac{u_1}{2\pi}$ uniformly on $\left[0, \frac{1}{2\pi}\right]$.

(3) Let $r = \sqrt{-2\ln u_1}$ and generate a random deviate $\theta = 2\pi u_2$, uniformly on $[0, 2\pi]$.

(4) Let $x_1 = r\cos\theta$ and $x_2 = r\sin\theta$. Then x_1 and x_2 are normal uncorrelated deviates.

In their paper, Box and Muller (1958) proved the validity of the method by reference to properties of the Chi-square (χ^2) distribution. However, we may give an alternative proof based on the following theorem (Troutt, 1991).

Theorem 1.1   *If a random variable X has a density $f(\mathbf{x})$, $\mathbf{x} \in \Re^n$, and if the random variable $v = f(\mathbf{x})$ has a density $g(v)$ then*

$$g(v) = -vA'(v) \quad . \tag{1.2}$$

where $A(v)$ is the Lebesgue measure of the set

$$S(v) = \{\mathbf{x} : f(\mathbf{x}) \geq v\}. \tag{1.3}$$

Remark: It is noted that a density for v need not exist. For example, consider the case in which $f(x)$ is uniformly distributed on $[0, 1]$.

Proof. Let m denote the maximum ordinate of the density $f(\mathbf{x})$. Then as suggested by Figure 1.3,

Proof. Let m denote the maximum ordinate of the density $f(\mathbf{x})$. Then as suggested by Figure 1.3, we calculate the cumulative distribution function (CDF) of $g(v)$:

$$
\begin{aligned}
G(v) &= \Pr(f(x) \le v) \\
&= 1 - \Pr(f(x) \ge v) \\
&= 1 - \int_{S(v)} f(x)\,\mathrm{d}x \\
&= 1 - vA(v) - \int_{v}^{m} A(s)\,\mathrm{d}s
\end{aligned}
\qquad (1.4)
$$

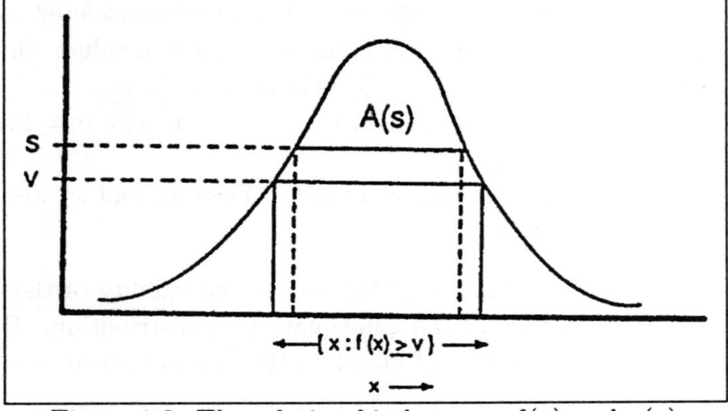

Figure 1.3: The relationship between $f(x)$ and $g(v)$

Hence by differentiation,

$$
g(v) = -vA'(v), \quad 0 \le v \le m. \qquad (1.5)
$$

The proof is complete.

Now we may verify for (1.1) that

$$
A(v) = -2\pi \ln(2\pi v),
$$

and so by (1.2), $g(v) = 2\pi$, $0 \le v \le \frac{1}{2\pi}$. That is, the ordinate of this density, considered as a random variable, is uniform, and thus an alternative (perhaps more intuitive) interpretation can be established as follows. A density value v is obtained according to the uniform density $f_{u_1}(u_1) = 2\pi$ on $[0, \frac{1}{2\pi}]$. Setting $v = 2\pi u_1$, there is a circle, $r^2 = -2\ln v$, associated with the height of that density. Lastly, a point $\mathbf{x} = (x_1, x_2)$ may be selected may be helpful as an alternative for obtaining $g(v)$. As an example, consider the standard normal density,

$$f(x) = \frac{1}{\sqrt{2\pi}} e^{-\frac{x^2}{2}}, \quad x \in \Re.$$

Then

$$A(v) = 2(-2\ln(v\sqrt{2\pi}))^{\frac{1}{2}},$$

and

$$A'(v) = -2v^{-1}(-2\ln(v\sqrt{2\pi}))^{-\frac{1}{2}},$$

so

$$g(v) = vA'(v) = 2(-2\ln(v\sqrt{2\pi}))^{\frac{1}{2}}.$$

For the multivariate normal pdf, with $x \in \Re^n$ and $f(\mathbf{x}) = (2\pi)^{-n/2}$ $\exp\{-\frac{1}{2}\mathbf{x}'\mathbf{x}\}$, we have $A(v) = \alpha_n[-2\ln\{(2\pi)^{n/2}v\}]^{n/2}$, where $\alpha_n = \pi^{n/2}$ $[(n/2)\Gamma(n/2)]^{-1}$ is the volume of the unit sphere in $\Re^n$ (see, for example, Fleming, 1977). Then $A'(v) = -(n/v)\alpha_n[-2\ln\{(2\pi)^{n/2}v\}]^{n/2-1}$ and the resulting vertical density is given by $g(v) = n\alpha_n[-2\ln\{(2\pi)^{n/2}v\}]^{n/2-1}$, for $0 \le v \le (2\pi)^{-n/2}$.

Figure 1.4 shows this $g(v)$-pdf for $n = 1, 2$ and 3. The reader may find it interesting to compare the ease of this approach with a more routine first principles one.

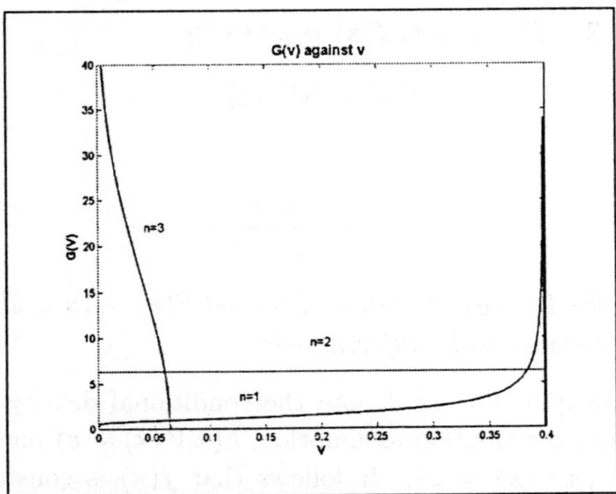

Figure 1.4: Graphs of $g(v)$ for $n = 1, n = 2$ and $n = 3$, respectively.

The Box-Muller method continues to be popular for generation of normal random deviates despite the existence of other methods which have been claimed to be more computationally efficient (see Marsaglia and Bray, 1964; Atkinson and Pearce, 1976; Ahrens and Dieter, 1972 and Kinderman and Ramage, 1976).

1.2.1 *A Formal Definition of Vertical Density Representation (VDR)*

As we have seen in Section 1.2, the idea of the alternative proof of the Box-Muller method was to first randomly generate an ordinate of the density; next solve for the equi-density contour associated with that ordinate; and finally randomly generate a point on that contour. This idea can be generalized for representing a large class of densities and gives rise to the following definition of vertical density representation (VDR).

Definition 1.1 Let $V(\mathbf{x})$ be a function on $\Re^n$ with range $[a,b]$. Let $g(v)$ be a given density on $[a,b]$. Define $S(v) = \{\mathbf{x} \in \Re^n : V(\mathbf{x}) \geq v\}$. Suppose for each $v \in [a,b]$, $\mathbf{x}$ is distributed uniformly on the boundary $\partial S(v)$. Then the process which selects $v \in [a,b]$ according to $g(v)$, and then $\mathbf{x} \in \partial S(v)$ according to a uniform density produces a random variable with a density $f(\mathbf{x})$ on $\Re^n$, is called the vertical density representation (VDR) and the density $g(v)$ is called the *vertical* or *ordinate density*.

Based on this definition, we have the following main result.

Theorem 1.2 *The density $f(\mathbf{x})$ is given by*

$$f(\mathbf{x}) = \phi(V(x)) \tag{1.6}$$

where

$$\phi(v) = -\frac{g(v)}{A'(v)} \tag{1.7}$$

and $A(v)$ is the Lebesgue measure of the set $S(v) = \{\mathbf{x} \in \Re^n : V(\mathbf{x}) \geq v\}$ and $A(v)$ is assumed to be differentiable.

Proof: Let $h(\mathbf{x}|V(\mathbf{x}) = v)$ denote the conditional density of $\mathbf{x}$ given v. By the uniform distribution assumption, $h(\mathbf{x}|V(\mathbf{x}) = v)$ must be constant on $\partial S(v) = \{\mathbf{x}|V(\mathbf{x}) = v\}$. It follows that $f(\mathbf{x})$ is constant on $\partial S(v)$, and hence, the level curves of $f(\mathbf{x})$ and $V(\mathbf{x})$ must coincide. Thus $f(\mathbf{x}) =$

$\phi(V(\mathbf{x}))$ for some $\phi(v)$. Let $G(v)$ denote the CDF for $g(v)$. Now for $v \in [a, b]$ and for all $\varepsilon > 0$ sufficiently small we have

$$\begin{aligned}
G(v + \varepsilon) - G(v) &= \Pr(\mathbf{x} \in \{\mathbf{x} : v \leq V(\mathbf{x}) \leq v + \varepsilon\}) \\
&= \int_{(\mathbf{x}: v \leq V(\mathbf{x}) \leq v + \varepsilon)} f(\mathbf{x}) \prod_{i=1}^{n} \mathrm{d}x_i \\
&\doteq \phi(v)[A(v) - A(v + \varepsilon)].
\end{aligned} \tag{1.8}$$

Then

$$\begin{aligned}
g(v) &= \lim_{\varepsilon \to 0} \frac{G(v + \varepsilon) - G(v)}{\varepsilon} \\
&= \phi(v)(-A'(v))
\end{aligned} \tag{1.9}$$

and hence the result.

Theorem 1.1 can be derived as a corollary of Theorem 1.2. To see this, note that with $V(\mathbf{x}) = f(\mathbf{x})$ in Theorem 1.1. ϕ is therefore the identity function. Thus (1.7) becomes $v = \frac{g(v)}{-A'(v)}$ and hence Theorem 1.1.

Remark: For many applications, the function $V(\mathbf{x})$ represents the distance of $\mathbf{x}$ from some target vector of interest, say, $\mathbf{x}^0$. For instance, $V(\mathbf{x})$ might be given as $(\mathbf{x} - \mathbf{x}^0)'Q(\mathbf{x} - \mathbf{x}^0)$ where Q is a real positive definite matrix. Here the definition $S(v) = \{\mathbf{x} : V(\mathbf{x}) \leq v\}$ becomes appropriate, since using the earlier definition would yield a set of unbounded measure. We continue to define $A(v) = L(S(v))$ where $L(\cdot)$ denotes Lebesgue measure. In this case, (1.7) becomes $\phi(v) = \frac{g(v)}{A'(v)}$.

1.3   Elementary Applications of Theorem 1.2

1.3.1   *VDR as a Kind of Variable Transformation*

VDR may be regarded as a special type of variable transformation. When the density of a variable of interest, $V(\mathbf{x})$, is desired, and a density $f(\mathbf{x})$ can be recognized as a function of $V(\mathbf{x})$, then Theorem 1.2 can be applied to find $g(v)$, the density of $V(\mathbf{x})$. An example is the quadratic form in the multivariate normal pdf on $\Re^n$ given by $f(\mathbf{x}) = 2\pi^{-n/2}|\Sigma|^{-1/2}\exp\{-1/2(\mathbf{x} - \mu)'\Sigma^{-1}(\mathbf{x} - \mu)\}$. Here, we obtain the well known result that the quadratic form $V(\mathbf{x}) = -1/2(\mathbf{x} - \mu)'\Sigma^{-1}(\mathbf{x} - \mu)$ has the gamma distribution $g(a, b)$, where $a = n/2$ and $b = 2$. The set $S(v) = \{\mathbf{x} : (\mathbf{x} - \mu)'\Sigma^{-1}(\mathbf{x} - \mu) \leq v\}$

is known to have Lebesgue measure $A(v)$, given by $A(v) = \alpha_n \Sigma^{-1} v^{n/2}$, where $\alpha_n = \frac{\pi^{n/2}}{n/2\Gamma(n/2)}$, is the volume of the unit sphere in $\Re^n$ (Fleming, 1977). It follows that $A'(v) = n/2\alpha_n|\Sigma|^{1/2}v^{n/2-1}$. Also, by inspection, $f(\mathbf{x}) = \phi(V(\mathbf{x}))$ implies that $\phi(v) = 2\pi^{-n/2}|\Sigma|^{-1/2}\exp\{-\frac{1}{2}v\}$. Hence by Theorem 1.2, we have $g(v) = \phi(v)A'(v) = (\Gamma(a)\beta^\alpha)^{-1}v^{\alpha-1}\exp\{-\frac{v}{\beta}\}$ where $a = n/2$ and $b = 2$. This distribution is also called the Chi-square (χ^2) distribution with n degrees of freedom (see, for example, Law and Kelton, 1982).

1.4   Construction of Hybrid Densities: Some Univariate Examples

A large number of density functions can easily be created by somewhat arbitrarily selecting $V(\mathbf{x})$ and $g(v)$, and then applying Theorem 1.2. Namely, $f(\mathbf{x}) = \phi(V(\mathbf{x}))$, where $\phi(v) = \pm\frac{g(v)}{A'(v)}$. In general, this will work provided the range of $V(\mathbf{x})$, $Ran(V)$, is the same as the support of $g(v)$. Several of these are presented in this section. Some are familiar and some are new. The function $V(\mathbf{x})$ may or may not be chosen as a pdf itself. For examples 1.1 - 1.4, $S(v)$ is defined by $S(v) = \{x : V(x) \geq v\}$, leading to the minus sign in the above $\phi(v)$-formula. For examples 1.5, 1.6 and 1.8, it is defined by $S(v) = \{x : V(x) \leq v\}$, which leads to the *plus* sign in that formula. These examples indicate the versatility of Theorem 1.2 in density modeling. Verification is left to the reader.

Examples in which V(x) itself is a pdf:

Example 1.1 $V(x) = \pi^{-1}(1+x^2)^{-1}, x \in (-\infty, \infty)$, $Ran(V) = [0, \pi^{-1}]$,
$$\text{(Cauchy pdf)}$$

$g(v) = \pi, v \in [0, \pi^{-1}]$,   (uniform pdf)
$A(v) = 2[(\pi v)^{-1} - 1]^{+1/2}$,
$A'(v) = \pi[(\pi v)^{-1} - 1]^{-1/2}[-(\pi v)^{-2}]$,
Result: $f(x) = |x|(1 + x^2)^{-2}, x \in (-\infty, \infty)$

Example 1.2
$V(x) = (2\pi)^{-1/2}\exp(-1/2x^2), x \in (-\infty, \infty)$, $Ran(V) = [0, (2\pi)^{-1/2}]$,
$$\text{(standard normal pdf)}$$
$g(v) = (2\pi)^{1/2}, v \in [0, (2\pi)^{-1/2}]$,   (uniform pdf)
$A(v) = 2(-2\ln[(2\pi)^{1/2}v])^{1/2}$,

$$A'(v) = -2v^{-1}(-2\ln[(2\pi)^{1/2}v])^{-1/2},$$
Result: $f(x) = \frac{1}{2}|x|\exp(-\frac{1}{2}x^2)$.

Example 1.3

$V(x) = \exp(-x), x \in [0, \infty), Ran(V) = (0, 1]$ (Laplace pdf)

$g(v) = \alpha[\exp(\alpha) - 1]^{-1}\exp(\alpha v), v \in [0, 1], \alpha > 0,$

$A(v) = -\ln v,$

$A'(v) = -v^{-1},$

Result: $f(x) = \alpha[\exp(\alpha) - 1]^{-1}\exp(-x)\exp(\alpha\exp(-x)), x \in (0, \infty).$

Example 1.4

$V(x) = \frac{1}{2}x^{-1/2}, x \in (0, 1), Ran(V) = (0, \infty),$

$g(v) = \exp(-v), v \in (0, \infty),$

$A(v) = \frac{1}{4}v^{-2}, A'(v) = -\frac{1}{2}v^{-3},$

Result: $f(x) = \frac{1}{4}x^{-3/2}\exp(-\frac{1}{2}x^{-1/2}), x \in (0, 1).$

Remark: The above result densities may humorously be called "Frankenstein" densities by analogy to the Frankenstein monster, which was supposed to have been constructed from miscellaneous parts of human bodies.

Examples in which $V(x)$ is not a pdf:

Example 1.5

$V(x) = x^2, x \in (-\infty, \infty), Ran(V) = [0, \infty)$

$g(v) = \exp(-v), v \in [0, \infty),$ (Laplace pdf)

$A(v) = 2v^{1/2}, A'(v) = v^{-1/2},$

Result: $f(x) = |x|\exp(-x^2), x \in (-\infty, \infty).$

Example 1.6

$V(x) = |x|, x \in (-\infty, \infty), Ran(V) = [0, \infty)$

$g(v) = \mu\exp(-\mu v), v \in [0, \infty), \mu > 0,$ (exponential pdf)

$A(v) = 2v, A'(v) = 2$

Result: $f(x) = \frac{1}{2}\mu\exp(-\mu|x|), x \in (-\infty, \infty).$ For $\mu = 1$, this is the double exponential density.

Example 1.7

$V(x) = \exp[-\frac{1}{2}x^2], x \in (-\infty, \infty), Ran(V) = [0, 1],$

$g(v) = \alpha[\exp(\alpha) - 1]^{-1}\exp(\alpha v), v \in [0, 1], \alpha > 0,$

$A(v) = 2(-2\ln v)^{1/2}, A'(v) = -2v^{-1}(-2\ln v)^{-1/2},$

Result: $f(x) = \frac{1}{2}\alpha[\exp(\alpha) - 1]^{-1}|x|[\exp(\frac{1}{2}x^2)][\exp\{\alpha\exp(-\frac{1}{2}x^2)\}], x \in (-\infty, \infty)$.

Example 1.8

$V(x) = (x - \mu)^2, x \in (-\infty, \infty), Ran(V) = [0, \infty),$

$g(v) = \alpha\beta^{-\alpha}v^{\alpha-1}\exp[-(\frac{v}{\beta})^\alpha], v \in [0, \infty),$ (Weibull pdf)

$A(v) = 2v^{\frac{1}{2}}, A'(v) = v^{-\frac{1}{2}},$

Result: $f(x) = \alpha\beta^{-\alpha}(x - \mu)^{2\alpha-1}\exp[-\{\frac{(x-\mu)^2}{\beta}\}^\alpha]$.

1.5   Further Results in VDR-Type Density Representation

In Section 1.2, we saw that if $v = f(\mathbf{x}) = \frac{1}{2\pi}\exp\{-\frac{1}{2}\mathbf{x}'\mathbf{x}\}$, $\mathbf{x} \in \Re^2$, then v has the uniform density on $[0, (2\pi)^{-1}]$. That is, the ordinate of the standard uncorrelated bivariate normal density is uniformly distributed. To generate a pair of independent $N(0, 1)$ variates we carry out the following two-stage procedure:

    (1) Randomly generate an ordinate of the density and solve for the isodensity contour associated with that ordinate;

    (2) Randomly generate a point on that contour.

The above idea can be extended to general univariate representation cases as follows (Troutt and Pang, 1997). For a given unimodal pdf $f(x)$ on $\Re$, we consider the set $S(v) = \{x : f(x) \geq v\}$. Let $h(x|v)$ be the conditional density on the boundary of $S(v)$. If $f(x)$ is unimodal, then $f(x) = v$ will have two solutions, namely $x^-(v)$ and $x^+(v)$, which coincide at the mode. We now write

$$h(x|v) = \frac{1}{2}\delta_{x^-(v)}(x) + \frac{1}{2}\delta_{x^+(v)}(x) \tag{1.10}$$

where $\delta_{x_0}(x)$ is the Dirac delta concentrated at x_0. Also let

$$m = \max\{f(x); x \in \Re\} \tag{1.11}$$

and $g(v)$, be the density of the density function $f(x)$ itself. Then

$$f(x) = \int_0^m h(x|v)g(v)\mathrm{d}x. \tag{1.12}$$

where $g(v)$ is the vertical or ordinate density.

It is noted that equation (1.10) and equation (1.11), as well as, $g(v)$ provide one particular solution of the general density composition or mixing representation in (1.12). Another particular solution pair can be derived from the Box-Muller method.

Suppose we consider

$$x = rw$$

where

$$w = \cos(2\pi u_2).$$

Clearly, the associated random variable W is distributed on the interval $[-1, +1]$. It can be easily checked that $f_W(w)$, the pdf of W, is given by

$$f_W(w) = \frac{1}{\pi} \frac{1}{\sqrt{1 - w^2}}, \quad w \in [-1, +1]. \tag{1.13}$$

Fig. 1.5 shows the graph of $f_W(w)$.

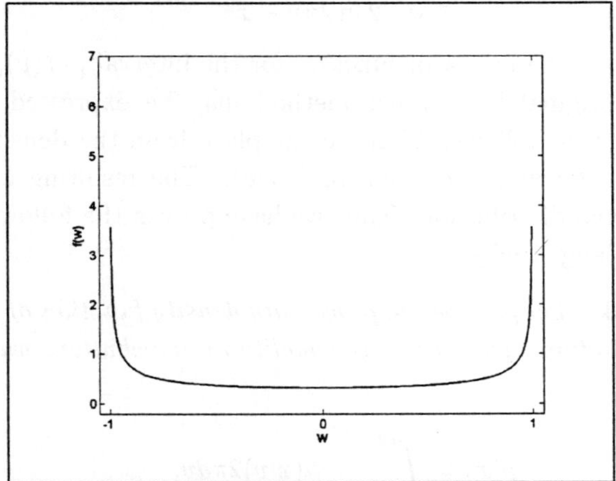

Figure 1.5: Graph of $f_W(w)$.

From Section (1.1.1), we note that

$$v = \frac{1}{2\pi} \exp\left\{ -\frac{1}{2} r^2 \right\}$$

and

$$r(v) = \sqrt{-2\ln 2\pi v}.$$

Let $x(v) = wr(v)$, $x(v) \in [r(v), r(v)]$ and consider $h(x|v)$, the density of $x(v)$. It is easy to show that

$$h(x|v) = \frac{1}{\pi r(v)} \frac{1}{\sqrt{1 - \left(\frac{x}{r(v)}\right)^2}}, \quad x \in [-r(v), r(v)], \tag{1.14}$$

where

$$r(v) = \sqrt{-2\ln(2\pi v)}$$

By substitution of $r(v)$ into $h(x|v)$, we obtain

$$h(x|v) = \frac{1}{\pi} \frac{1}{\sqrt{-2\ln 2\pi v - x^2}}, \quad x \in [-r(v), r(v)], \tag{1.15}$$

or

$$h(x|v) = \frac{1}{\pi} \frac{1}{\sqrt{-2\ln 2\pi v - x^2}} I_{r(v)}(x), \tag{1.16}$$

where $I_{r(v)}(x)$ is the indicator function for the interval $[-r(v), r(v)]$.

Now the original Box-Muller method may be expressed in terms of $h(x|v)$ and $g(v)$ as follows. First we sample v from the density $g(v)$, and then sample x given v according to $h(x|v)$. The resulting x follows the standard normal distribution. Thus we have proven the following theorem (Troutt and Pang, 1997).

Theorem 1.3 *Let $p(x)$ be the probability density function of the standard normal distribution. Then we have equality in distribution, which we write as follows:*

$$p(x) = \int_0^{(2\pi)^{-1}} h(x|v) 2\pi dv, \tag{1.17}$$

where $h(x|v) = \frac{1}{\pi} \frac{1}{\sqrt{-2\ln 2\pi v - x^2}} I_{r(v)}(x)$ and $I_{r(v)}(x)$ is the indicator function for the interval $[-(-2\ln 2\pi v)^{\frac{1}{2}}, (-2\ln 2\pi v)^{\frac{1}{2}}]$.

The above procedure is a compositional Monte Carlo method for the standard normal density. The method should clearly extend to a large class of bell-shaped distributions except that $g(v)$ will not in general be the uniform pdf. The difference between the VDR method proposed by Troutt (1993) and the method in this section may be described as follows. The original VDR approach only generates a point on the boundary of the set $S(v)$, while the method in this section generates a point from the entire set $S(v)$. **Remark.** In fact, equation (1.17) actually holds in the usual algebraic sense as well. The reader may find it instructive to verify this using the substitution, $y^2 = -\ln(2\pi v) - x^2$. An extension of Theorem 1.3 is given in Section 3.3. We may note one further solution of the composition equation (1.12). First, we observe that $\int f(x)\mathrm{d}x = 1 = \int A(v)\mathrm{d}v$, where $A(v) = L\{S(v)\} = L\{x : f(x) \geq v\}$. Thus, since also $A(v) \geq 0, A(v)$ is itself a pdf with support equal to the range of $f(x)$, say, $[0, m]$. If $A(v)$ is strictly positive and bounded, we define

$$h(x|v) = \frac{1}{A(v)}I_{S(v)}.$$

Noting that $h(x|v) = 0$ if $v > f(x)$, we then obtain

$$\int_0^m h(x|v)A(v)\mathrm{d}v = \int_0^{f(x)} \mathrm{d}v = f(x)$$

This may be regarded as an intuitive idea similar to Type II VDR first discussed by Fang *et al.* (2001), which is discussed in detail in Sections 3.1 and 4.5. In addition, this can be viewed from the perspective of the acceptance-rejection (AR) method from simulation theory, which is discussed further in Chapter 4. That is, if for given v where $0 \leq v \leq f(x)$, $f(x)$ has the conditional density $h(v|x) = 1/f(x)$, then clearly the joint pdf of x and v on $\{(x, v) : x \in (-\infty, \infty), 0 \leq v \leq f(x)\}$ is uniform under the assumption of independence. Further details on this connection are given in Section 4.6.

1.6 Further Remarks on Vertical Density Representation

As mentioned in the previous section, the concept of vertical density representation can be used to represent a large class of densities. Previous work on VDR assumed that the conditional density of the variate on isodensity contours is uniform. In this section, we show the results for more general contour densities.

The conditional density $h(x|v)$ which we also call the contour density, plays a critical role in VDR (Troutt, 1991; Troutt, 1993; Troutt and Pang, 1997). Let $\mathbf{x} \in \Re^n$ and let $V(\mathbf{x})$ be a real valued function on $\Re^n$ such that (i) the range of $V(\mathbf{x})$ is $[0, \infty)$, and (ii) for each $\mathbf{x}_0 \in \Re^n$, there exists a $v_0 \in [0, \infty)$ with $V(\mathbf{x}_0) = v_0$. The set $\{\mathbf{x} : V(\mathbf{x}) = v_0\}$ is a level curve or contour of $V(\mathbf{x})$. There are two questions regarding densities in $\Re^n$ in this context. First let a family of densities on the contours be given. Namely, let $h(\mathbf{x}|v)$, called a contour density, be a probability density function on $\{\mathbf{x} : V(\mathbf{x}) = v\}$ for each $v \in [0, \infty)$. Finally, let $g(v)$ be a probability density function on $[0, \infty)$.

The first question is as follows. Suppose that realizations, v, of $v \in [0, \infty)$ are generated according to the pdf $g(v)$. Then given the value of v, a vector $\mathbf{x} \in \{\mathbf{x} : V(\mathbf{x}) = v\}$ is generated according to the contour pdf $h(\mathbf{x}|v)$. What will be the pdf of the $\mathbf{x}$-realizations on $\Re^n$? Conversely, given a pdf $f(\mathbf{x})$ on $\Re^n$ and a $V(\mathbf{x})$ of the above form, what are the corresponding $g(v)$ and $h(\mathbf{x}|v)$? These questions are clearly answered by specifying the relationship among $V(\mathbf{x})$, $g(v)$, $h(\mathbf{x}|v)$, and $f(\mathbf{x})$.

The foregoing relationship has so far been developed only for the assumption that $h(\mathbf{x}|v)$ is uniform in $\mathbf{x}$ for each respective value of v. A consequence of this assumption is that the set $\{\mathbf{x} : V(x) = v\}$ must be compact for each value of v.

Interest in these issues arises from several sources. Such representations provide new avenues for simulation purposes. Also, these representations generalize the idea of L_p-norm symmetric densities. They have also been applied to study the tail behavior of univariate densities. Finally, if $V(\mathbf{x})$ is a performance measure, then VDR enables inferences about the density of $V(\mathbf{x})$ scores from statistical observations related to $f(\mathbf{x})$.

1.6.1 *Nonuniform Contour Densities*

Recall that the result in Theorem 1.2 can be used to represent a large class of densities in the following manner. Let $V(\mathbf{x})$ be a function on $\Re^n$ with range $[a, b]$ and let $g(v)$ be a given density on $[a, b]$. Suppose for each $v \in [a, b]$, $\mathbf{x}$ is distributed uniformly on $\partial S(v) = \{\mathbf{x} \in \Re^n : V(\mathbf{x}) = v\}$, then the process which selects $v \in [a, b]$ according to $g(v)$, and then $\mathbf{x} \in \partial S(v)$ according to a uniform density produces a random variable with a density $f(\mathbf{x})$ on $\Re^n$. Also it was shown that the vertical density, $g(v)$, is related to $f(\mathbf{x})$ as follows. Theorem 1.2 may be interpreted to say that if the density

$f(\mathbf{x})$ is given by

$$f(\mathbf{x}) = \phi(V(\mathbf{x}))$$

for some real valued function $\phi : [a, b] \to \Re$, then

$$g(v) = -\phi(v)\frac{d}{dv}A(v), \qquad (1.18)$$

where $A(v)$ is the Lebesgue measure of the set $S(v)$.

The relation (1.18) above is a basic property of VDR. The aim of this section is to give a more general form for this particular relation. In fact, using a special change of variables formula for integrals, we prove in Section 1.7 the following main result, which expresses an explicit relationship among $V(\mathbf{x})$, $g(v)$ and $f(\mathbf{x})$:

$$g(v) = \int\limits_{\{\mathbf{x}:V(\mathbf{x})=v\}} \frac{f(\mathbf{x})}{\| \nabla V(\mathbf{x}) \|} \, d\sigma, \qquad (1.19)$$

where σ is the canonical measure on the level set $\{\mathbf{x} : V(\mathbf{x}) = v\}$. Formula (1.19) will facilitate the calculation of $g(v)$ given $f(\mathbf{x})$ and $V(\mathbf{x})$. Some examples are given in the next section.

1.7 Vertical Density Representation in the General Case

Let $(\Omega, \mathcal{A}, P)$ be a probability space and let the random vector

$$\mathbf{X} = (X_1, \cdots, X_n) : \Omega \to \Re^n \qquad (n \geq 2)$$

have a pdf, $f(\mathbf{x}), \mathbf{x} = (x_1, \cdots, x_n) \in \Re^n$. Let $V : \Re^n \to \Re$ be a real valued function with range $[a, b]$. Then

$$\Re^n = \bigcup_{a \leq v \leq b} \{\mathbf{x} : V(\mathbf{x}) = v\},$$

so that the n-dimensional Euclidean space, $\Re^n$, is the union of the level sets of the function V. In the sequel, we shall assume that the gradient vector ∇V vanishes nowhere on $\Re^n$, i.e.,

$$\nabla V(\mathbf{x}) \neq \mathbf{0}, \quad \forall \mathbf{x} \in \Re^n.$$

Thus, the level set $\{V = v\}$ is an oriented $(n-1)$-surface in $\Re^n$. It is tacitly assumed that $\mathbf{n} = \frac{\nabla V}{\|\nabla V\|}$ is the unit normal on the surface.

Theorem 1.4   *The pdf of $V(\mathbf{X})$ is given by*

$$g(v) = \int\limits_{\{V=v\}} \frac{f(\mathbf{x})}{\parallel \nabla V(\mathbf{x}) \parallel}\, d\sigma, \qquad v \in [a, b],$$

where σ is the canonical measure on the oriented $(n-1)$-surface $\{V = v\}$ with the unit outward normal $\mathbf{n} = \frac{\nabla V}{\parallel \nabla V \parallel}$.

Proof. The proof of Theorem 1.4 makes use of the following special change of variables formula $\displaystyle \int\limits_{\{v_1 \leq V \leq v_2\}} f(\mathbf{x})d\mathbf{x} = \int_{v_1}^{v_2} \left(\int\limits_{\{V=v\}} \frac{f(\mathbf{x})}{\parallel \nabla V(\mathbf{x}) \parallel}\, d\sigma \right) dv.$

Indeed, on the level set $S = \{\mathbf{x} : V(\mathbf{x}) = r\}$, the canonical measure σ is related to the volume form dS on S via

$$\sigma(A) = \int_A dS \qquad \text{for } A \subset \Re^n$$

Then by the theory of differential forms, we have

$$dS \quad = \frac{\nabla V}{\parallel \nabla V \parallel}\, dx_1 \wedge \cdots \wedge dx_n$$

$$= \frac{1}{\parallel \nabla V \parallel} \sum_{i=1}^{n} (-1)^{i-1} \frac{\partial V}{\partial x_i} dx_1 \wedge \cdots \wedge \cdots \wedge dx_n$$

Since $dv = \frac{\partial V}{\partial x_i} dx_1 + \cdots + \frac{\partial V}{\partial x_n} dx_n$ for $v = V(\mathbf{x})$, we have

$$dv \wedge dS \quad = \frac{1}{\parallel \nabla V \parallel} \cdot \parallel \nabla V \parallel^2 dx_1 \wedge \cdots \wedge dx_n$$

$$= \parallel \nabla V \parallel dx_1 \wedge \cdots \wedge dx_n,$$

so that

$$dx_1 \wedge \cdots \wedge dx_n = \frac{1}{\parallel \nabla V \parallel} dv \wedge dS.$$

We can now assert that

$$\int\limits_{\{v_1 \leq V \leq v_2\}} f(\mathbf{x})d\mathbf{x} = \int_{v_1}^{v_2} \left(\int\limits_{\{V=v\}} \frac{f(\mathbf{x})}{\parallel \nabla V(\mathbf{x}) \parallel}\, d\sigma \right) dv. \qquad (1.20)$$

Proof of Theorem 1.4. Using 1.20, we have

$$P\{v \leq V(\mathbf{X}) \leq v + \Delta\} = \int_{\{v \leq V \leq v+\Delta\}} f(\mathbf{x})d\mathbf{x}$$

$$= \int_v^{v+\Delta} \left(\int_{\{V=v\}} \frac{f(\mathbf{x})}{\| \nabla V(\mathbf{x}) \|} d\sigma \right) dv.$$

Thus,

$$g(v) = \lim_{\Delta \to 0} \frac{P\{v \leq V \leq v + \Delta\}}{\Delta} = \int_{\{V=v\}} \frac{f(\mathbf{x})}{\| \nabla V(\mathbf{x}) \|} d\sigma.$$

This completes the proof.

The original basic result in Theorem 1.1 is now an easy consequence of Theorem 1.4. Indeed, if $f(\mathbf{x}) = \phi(V(\mathbf{x}))$ is given, then we have

$$g(v) = \int_{\{V=v\}} \frac{f(\mathbf{x})}{\| \nabla V(\mathbf{x}) \|} d\sigma = \phi(v) \left(\int_{\{V=v\}} \frac{1}{\| \nabla V(\mathbf{x}) \|} d\sigma \right).$$

On the other hand,

$$A(v) = \int_{S(v)} d\mathbf{x} = \int_v^b \left(\int_{\{V=v\}} \frac{1}{\| \nabla V(\mathbf{x}) \|} d\sigma \right) dt,$$

so that

$$\frac{d}{dv} A(v) = - \int_{\{V=v\}} \frac{1}{\| \nabla V(\mathbf{x}) \|} d\sigma.$$

Hence,

$$g(v) = \phi(v) \left(-\frac{d}{dv} A(v) \right)$$

as claimed.

Theorem 1.5 *Let $h(\mathbf{x} \mid V = v)$ be the conditional density of $\mathbf{X}$ given $V = v$. Then*

$$f(\mathbf{x}) = g(v) \| \nabla V(\mathbf{x}) \| h(\mathbf{x} \mid V = v)$$

on $\{V = v\}$. In particular,

$$h(\mathbf{x} \mid V = v) = \frac{\frac{f(\mathbf{x})}{\|\nabla V(\mathbf{x})\|}}{\int\limits_{\{V=v\}} \frac{f(\mathbf{x})}{\|\nabla V(\mathbf{x})\|} \, d\sigma} \, \delta_{\{V=v\}}(\mathbf{x}),$$

where

$$\delta_{\{V=v\}}(\mathbf{x}) = \begin{cases} 1 & if \quad V(\mathbf{x}) = v \\ 0 & if \quad V(\mathbf{x}) \neq v. \end{cases}$$

Proof. Let I_A be the indicator function of a Borel set A in $\Re^n$. Then

$$\int_a^b \left(\int\limits_{\{V=v\}} I_A(\mathbf{x}) h(\mathbf{x} \mid V = v) \, d\sigma \right) g(v) dv$$

$$= \int\limits_{\{a \leq V \leq b\}} I_A(\mathbf{x}) f(\mathbf{x}) \, d\mathbf{x}$$

$$= \int_a^b \left(\int\limits_{\{V=v\}} \frac{I_A(\mathbf{x}) f(\mathbf{x})}{\| \nabla V(\mathbf{x}) \|} \, d\sigma \right) dv$$

$$= \int_a^b \left(\int\limits_{\{V=v\}} \frac{I_A(\mathbf{x}) f(\mathbf{x})}{g(v) \| \nabla V(\mathbf{x}) \|} d\sigma \right) g(v) dv.$$

Thus,

$$\begin{aligned} h(\mathbf{x} \mid V = v) \;&=\; \frac{f(\mathbf{x})}{g(v) \| \nabla V(\mathbf{x}) \|} \qquad \text{on } \{V = v\} \\[2ex] &=\; \frac{f(\mathbf{x})}{g(v) \| \nabla V(\mathbf{x}) \|} \delta_{\{V=v\}}(\mathbf{x}). \end{aligned}$$

This completes the proof.

1.7.1 *Examples*

Example 1.9 This example was given in Troutt and Pang (1997).

Let $\mathbf{x} = (x, y)$ and $f(\mathbf{x}) = f(x, y) = \frac{1}{2\pi} e^{-\frac{r^2}{2}}$ where $r = \sqrt{x^2 + y^2} = \| \mathbf{x} \|$. Then

$$f(\mathbf{x}) = \phi(V(\mathbf{x}))$$

with $V(\mathbf{x}) = \| \mathbf{x} \|^2$ and $\phi(t) = \frac{1}{2\pi} e^{-\frac{t}{2}}$. Since $\nabla V = (2x, 2y)$, we have $\| \nabla V(\mathbf{x}) \| = 2r = 2\sqrt{V}$. Thus, on $V = v$, $d\sigma = \sqrt{v}d\theta$, $0 \le \theta \le 2\pi$

$$
\begin{aligned}
g(v) \quad &= \int_{\{V=v\}} \frac{f(\mathbf{x})}{\| \nabla V(\mathbf{x}) \|} d\sigma \\
&= \phi(v) \int_0^{2\pi} \frac{1}{2\sqrt{v}} (\sqrt{v}d\theta) \\
&= \phi(v)\pi \\
&= \frac{e^{-\frac{v}{2}}}{2} \quad \text{on} \quad [0, \infty].
\end{aligned}
$$

On the other hand, if we let $V(\mathbf{x}) = f(\mathbf{x})$

$$
\begin{aligned}
\nabla f \quad &= \left(\frac{1}{2\pi} e^{-\frac{r^2}{2}} (-x), \; \frac{1}{2\pi} e^{-\frac{r^2}{2}} (-y) \right) \\
&= -\frac{1}{2\pi} e^{-\frac{r^2}{2}} \mathbf{x}
\end{aligned}
$$

so that

$$
\begin{aligned}
\| \nabla f \| \quad &= \frac{1}{2\pi} e^{-\frac{r^2}{2}} r \\
&= f(\mathbf{x}) r
\end{aligned}
$$

Therefore

$$
\begin{aligned}
g(v) \quad &= \int_{\{f=v\}} \frac{f(\mathbf{x})}{\| \nabla f(\mathbf{x}) \|} r d\theta \\
&= 2\pi \quad \text{on } [0, \; (2\pi)^{-1}].
\end{aligned}
$$

The level set, $\{\mathbf{x} : f(\mathbf{x}) = v\}$, is a circle of radius $r(v) = \sqrt{(-2\ln 2\pi v)}$ and

$$h(\mathbf{x} \mid f = v) = \frac{1}{2\pi r(v)} \delta_{\{f=v\}}(\mathbf{x})$$

Thus, $f(\mathbf{x}) = \int_0^{\frac{1}{2\pi}} h(\mathbf{x}|f = v)g(v)dv$. Note that $h(\mathbf{x}|f = v)$ is not a smooth function on the whole space. In fact, it is degenerate.

Example 1.10 Let $f(\mathbf{x}) = w(\parallel \mathbf{x} \parallel^2)$, $\mathbf{x} \in \Re^n$ where $w(\cdot)$ is a strictly decreasing, differentiable and one-to-one real function. It is easy to calculate that on $f(\mathbf{x}) = v$,

$$\parallel \nabla f(\mathbf{x}) \parallel^2 = 4(w'[w^{-1}(v)])^2 w^{-1}(v).$$

Thus the pdf, $g(v)$ of $V = f(\mathbf{x})$ is given by

$$g(v) \quad = \quad \int_{\{f=v\}} \frac{f(\mathbf{x})}{\parallel \nabla V(\mathbf{x}) \parallel} d\sigma$$

$$= \frac{-v}{2w'[w^{-1}(v)]\sqrt{w^{-1}(v)}} A_{n-1}$$

where A_{n-1} is the surface area of the sphere of radius $\sqrt{w^{-1}(v)}$ given by

$$A_{n-1} = \frac{n\pi^{\frac{n}{2}}}{\Gamma(\frac{n}{2}+1)}.$$

Therefore

$$g(v) = \frac{-n\pi^{\frac{n}{2}} v}{2\Gamma(\frac{n}{2}+1)w'[w^{-1}(v)]\sqrt{w^{-1}(v)}}.$$

Example 1.11 Let the random vector $\mathbf{X} = (X_1, \ldots, X_n) \in \Re^n_+$ have the pdf, $f(\mathbf{x}) = h(\parallel \mathbf{x} \parallel_1)$, where $\parallel \mathbf{x} \parallel_1 = \sum_{i=1}^n \mid x_i \mid = |x_1| + \cdots + |x_n|$, and $h(\cdot)$ be a strictly decreasing and differentiable function. Then on $f = v$,

$$\nabla f(\mathbf{x}) = h'[h^{-1}(v)](1, 1, \ldots, 1)^t,$$

so that

$$\parallel \nabla V(\mathbf{x}) \parallel = -h'[h^{-1}(v)]\sqrt{n}.$$

Thus,

$$g(v) \quad = \quad \int_{\{f=v\}} \frac{f(\mathbf{x})}{\parallel \nabla V(\mathbf{x}) \parallel} d\sigma$$

$$= \frac{-v}{h'[h^{-1}(v)]\sqrt{n}} \int_D d\sigma,$$

where $D = \{(x_1, \ldots, x_n) \mid x_1 + \cdots + x_n = h^{-1}(v), \ x_1 \geq 0, \ldots, x_n \geq 0\}$.

1.8 Comments

As noted earlier initial work with the VDR concept assumed a uniform conditional distribution $h(\mathbf{x} \mid V = v)$ for each value of v. In Section 1.4 results for the general nonuniform conditional or contour density case were obtained. These results suggest that given a pdf $f(\mathbf{x})$ and a system of contours generated by a function $V(\mathbf{x})$ there exist densities, $g(v)$ and $h(\mathbf{x} \mid V = v)$ so that the relationships of Theorem 1.5 hold among the four components. Thus a wide variety of different representations of $f(\mathbf{x})$ become possible as the function $V(\mathbf{x})$ is varied. Alternative representations may be useful in developing Monte Carlo simulation techniques.

Interest in such representations also arises in the analysis of performance. For example, if $V(\mathbf{x}) = (\mathbf{x} - \mathbf{x}^0)'Q(\mathbf{x} - \mathbf{x}^0)$ is taken as the performance measure for a shooter firing at a target located at position $\mathbf{x} = \mathbf{x}^0$ then $f(\mathbf{x})$ is the spatial density for the location of hits and $g(v)$ is the density of the performance measure itself. In this setting, the density h might be expected to measure effects related to the relative orientations of the target and shooter.

Chapter 2

Applications of Vertical Density Representation

Introduction

This chapter presents two applications. The first is an application to the tail behaviour of pdfs and the ordering of their distributions. The discussion builds on the paper of Kotz and Troutt (1996) and adds some new results. The second application is to the analysis or decomposition of correlation into two components called *vertical* and *contour correlation*, respectively. An application to the aggregation of experts is discussed. This material has not been published previously.

2.1   Application I: Ordering of Distributions

We note that a pdf, $f(x)$, may not be uniquely defined by its vertical pdf, $g(v)$. For instance, the vertical pdf remains the same if the values of $f(x)$ over any interval are interchanged with those of another interval of equal length. The reader may verify that such interchanges do not change the values of the $A(v)$-function. In fact, $g(v)$ is unchanged under any horizontal deformation of the graph of $f(x)$ that continues to define a pdf and which does not change the existence or values of $A'(v)$. This follows from $g(v) = -vA'(v)$ in Theorem 1.1. Such changes may be called *strata shifts* and are applied in Chapter 5 and are also discussed further in Section 6.5. The resulting density may then be called a *strata shift density*. Also, the vertical pdfs of X and $X + \theta$ are the same. However, the vertical pdfs will differ for X and cX for $|c| \neq 0$ or 1. An approach that normalizes this last kind of difference is proposed next. We have omitted the case of

$c = 0$ because the concept of vertical pdf has so far not been defined for point-mass distributions or degenerate density functions.

For bounded densities, Kotz and Troutt (1996) define a new random variable, W with pdf, $p_W(w)$, $0 < w < 1$, $w = v/m$, $v = f(x)$ and $m = \max\{f(x)\}$. In particular, if $p_v(v) = g(v)$ then by a simple change of variables, we have $p_W(w) = mg(mw)$. The density, $p_W(w)$ may be called the *normalized* vertical pdf and $p_W(w)$ is the same for X and cX, $c \neq 0$. The proof of this claim illustrates basic properties of these constructs and is an application of Theorem 1.1. We therefore give it formally as follows.

Theorem 2.1   *Let random variable X have a bounded pdf, $f_X(x)$, for $x \in \Re$, with vertical density, $g_X(v), v \geq 0$. Let $Y = cX$ with $c \neq 0$. Then X and Y have the same normalized vertical density, $p_W(w)$.*

Proof: Let, $m = \max\{f_X(x) : x \in \Re\} > 0$ and let $w = v/m$. It is elementary that $f_Y(y) = |c|^{-1} f_X(c^{-1}y)$ and similarly for $W, p_W(w) = mw(g_X(mw))$, noting that $m > 0$. Let $A_X(v) = L\{x : f_X(x) \geq v\}$. Then

$$
\begin{aligned}
A_Y(v) &= L\{y : f_Y(y) \geq v\} \\
&= L\{y : y = cx \text{ and } |c|^{-1} f_X(c^{-1}y) \geq v\} \\
&= |c|L\{x : f_X(x) \geq |c|v\} \\
&= |c|A_X(|c|v).
\end{aligned}
$$

Let $g_Y(v)$ be the vertical pdf for Y. It follows from Theorem 1.1 that

$$
\begin{aligned}
g_Y(v) &= -v\frac{d}{dv}A_Y(v) \\
&= -v|c|^2 A'_X(|c|v) \\
&= |c|(-|c|vA'_X(|c|v)) \\
&= |c|g_X(|c|v).
\end{aligned}
$$

Since $\max\{f_Y(y) : y \in \Re\} = m/|c|$, then for Y,

$$
\begin{aligned}
p_W(w) &= [mw/|c|]g_Y(mw/|c|) \\
&= [mw/|c|][|c|g_X(|c|mw/|c|)] \\
&= mw(g_X(mw)),
\end{aligned}
$$

which is the same as $p_W(w)$ for X.

Remark: We may note that the fact, $g_Y(v) = |c|g_X(|c|v)$, does not depend on $f_X(x)$ being bounded.

Either $p_V(v)$ or $p_W(w)$ can be considered as an indicator of the tail behaviour of $f(x)$. We would expect both to be more concentrated at values near zero for relatively thick-tailed distributions. This idea is used further in Section 2.3 below.

2.2 Standard Measurement in Tail Behaviour Using VDR

Using the above argument, we may now establish a measurement of tail behaviours for some useful probability distributions (Kotz and Troutt, 1996). To start with, we consider the exponential and Laplace distributions. The two distributions have the same normalized vertical density, namely,

$$p_W(w) = \begin{cases} 1, & 0 < w < 1, \\ 0, & \text{elsewhere,} \end{cases} \tag{2.1}$$

which is uniform. It may serve as the measurement or etalon for the rate of tail ordinate decline for a distribution. We have $p_W(w) = 0$ for the standard uniform distribution. That is to say that the tails are constant with a measurement rate of zero, whereas, the class of power function distributions with pdf $p_X(x) = ax^{a-1}$, $0 < x < 1$, $a < 1$, have a continuously increasing vertical density from 0 to 1.

Kotz and Troutt (1996) provided a table that summarizes the V and W distributions for 13 parent distributions in the order of decreasing tail behaviour. Interested readers may refer to Table 1 in the article of Kotz and Troutt (1996) for full details. In the following we present the relationships between $f(x)$, $A(v)$, $p_V(v)$ and $p_W(w)$ for six common distributions. We limit our attention to densities that are strictly positive on either $[0, \infty)$ or $(-\infty, \infty)$.

(1) Pearson Type VII Distribution $(a > 1)$

$$f_X(x) = \frac{\Gamma(a)\sqrt{\pi}}{\Gamma(a-\frac{1}{2})}(1 + x^2)^{-a}, \quad x \in \Re; a > 0;$$

$$A(v) = 2\{(K_a/v)^{\frac{1}{a}} - 1\}^{\frac{1}{2}};$$

$$p_V(v) = a^{-1}(K_a/v)^{\frac{1}{a}}\{(K_a/v)^{\frac{1}{2}} - 1\}^{-\frac{1}{2}}, \quad 0 < v < K_a;$$

$$p_W(w) = a^{-1} K_a w^{-\frac{1}{a}} (w^{-\frac{1}{a}} - 1)^{-\frac{1}{2}},$$

where $K_a = \dfrac{\Gamma(a)\sqrt{\pi}}{\Gamma(a-\frac{1}{2})}$.

(2) Exponential Distribution

$$f_X(x) = e^{-x}, \quad x > 0;$$

$$A(v) = -\ln v;$$

$$p_V(v) = 1, \quad 0 < v < 1;$$

$$p_W(w) = 1.$$

(3) Pareto Distribution

$$f_X(x) = a k^a x^{-(a+1)}, \quad x > k > a > 0;$$

$$A(v) = \left(\frac{a k^a}{v}\right)^{\frac{1}{(a+1)}} - k;$$

$$p_V(v) = (a+1)^{-1}\left(\frac{a k^a}{v}\right)^{\frac{1}{(a+1)}}, \quad 0 < v < \frac{a}{k};$$

$$p_W(w) = a(a+1)^{-1} w^{-\frac{1}{(a+1)}}.$$

(4) Normal Distribution

$$f_X(x) = \frac{1}{\sqrt{2\pi}} e^{-\frac{x^2}{2}}, \quad x \in \Re;$$

$$A(v) = 2\{-2\ln(v\sqrt{2\pi})\}^{\frac{1}{2}};$$

$$p_V(v) = 2\{-2\ln(v\sqrt{2\pi})\}^{-\frac{1}{2}}, \quad 0 < v < (\sqrt{2\pi})^{-1};$$

$$p_W(w) = \sqrt{\frac{2}{\pi}}(-2\ln w)^{-\frac{1}{2}}.$$

(5) Cauchy Distribution

$$f_X(x) = \frac{1}{\pi} \frac{1}{(1+x^2)}, \quad x \in \Re;$$

$$A(v) = 2\{(\pi v)^{-1} - 1\}^{\frac{1}{2}};$$

$$p_V(v) = \frac{1}{\sqrt{\pi v(1 - \pi v)}}, \quad 0 < v < \frac{1}{\pi};$$

$$p_W(w) = \frac{1}{\pi} \frac{1}{\sqrt{w(1-w)}},$$

which is the arcsine density.

(6) Logistic Distribution

$$f_X(x) = \frac{e^x}{(e^x + 1)^2}, \quad x \in \Re;$$

$$A(v) = 2\ln\left(\frac{1 - 2v + \sqrt{1 - 4v}}{2v}\right);$$

$$p_V(v) = \frac{2}{\sqrt{1 - 4v}}, \quad 0 < v < \frac{1}{4};$$

$$p_W(w) = \frac{1}{2} \frac{1}{\sqrt{(1-w)}}.$$

2.2.1 *Discussion*

The vertical pdf provides a transparent indicator of certain aspects of the tail behavior of univariate distributions. However, it may not be possible to obtain an exact solution (for x) of the inequality $p_X(x) > v$. In such cases, we may not be assured of explicit forms for $A(v)$ or $A'(v)$. We can, in principle, obtain $A'(v)$ by numerical methods. For example, consider the Weibull $p_X(x) = cx^{c-1}\exp(-x^c)$, where calculation of $p_X(x) > v$ implies $(c-1)\ln(x-x^c) > \ln(v/c)$. A similar case is typified by a Gamma-type distribution $p_X(x) = x\exp(-1/2x^2)\,(x > 0)$ where the calculation of $p_X(x)$ implies that $\ln x - 1/2x^2 > \ln v$ is required. Further interesting extension work on this topic would be the multivariate case. This will likely relate to the concept of a scalar-valued multivariate hazard rate (Goodman and Kotz, 1981).

2.3   A New Class of Very Thick-Tailed Densities

Next, we derive a new class of thick-tailed densities in an example, which also illustrates a different kind of application for Theorem 1.1. In that theorem, a pdf $f(x)$ was given and it was desired to find the associated vertical density $g(v)$. Here, we start with $g(v)$ and seek to determine $f(x)$. The new class, which we call the *negative power* vertical density class, has thicker tails than that of the Cauchy density.

Example 2.1   The negative power vertical density class

Consider selecting a vertical density from the class given by $g(v) = Cv^{-q}$ where $0 < q < 1, 0 < v < m$ and $m > 0$. Here, we may check that $C = C(m,q) = (1-q)m^{q-1}$. This class is unbounded at zero, the more so as q increases to 1.0. Hence, if we can find a symmetric $f(x)$ on $(-\infty, \infty)$ for which $g(v)$ is the vertical density then that $f(x)$ will have thick tails. From Theorem 1.1, we have that if $g(v)$ is the vertical density for $f(x)$, then we must have $g(v) = -vA'(v)$ where $v = f(x)$. We will solve for $A(v)$ in this relationship and then construct an $f(x)$ from that solution. This requires that $m = \max f(x)$. Then we have the differential equation,

$$A'(v) = -g(v)/v, \quad \text{with} \quad A(m) = 0, \tag{2.2}$$

To simplify, we will construct a solution that is unimodal and symmetric about zero. The reader may verify that problem (2.2) has the solution

$$A(v) = \frac{(1-q)}{q}m^{-q}(v^{-q} - m^{-q}), \tag{2.3}$$

Now, if we consider a horizontal line segment of length $A(v)$ with center on the vertical axis and positioned at height v, then the endpoints of the segment correspond to two x-values on the horizontal axis. These are given by $x^+(v)$ and $x^-(v)$. That is,

$$|x(v)| = \frac{1}{2}A(v) = \frac{1}{2q}(1-q)m^{-q}(v^{-q} - m^{-q}), \tag{2.4}$$

Solving for $v = f(x)$, and we obtain

$$f(x) = f(x; m, q) = (m^{-q} + 2q(1-q)m^{1-q}|x|)^{-\frac{1}{q}}, m > 0, 0 < q < 1. \quad (2.5)$$

One may verify by direct integration that $f(x)$ is a density for all admissible values of the parameters. For this class, we have $p_W(w) = (1-q)w^{-q}, w \in (0,1)$. For a specific case, let $m = 1$ and $q = \frac{3}{4}$. Then $f(x) = (1+(\frac{3}{8})|x|)^{-\frac{4}{3}}$. Fig. 2.1 shows the graphs of $f(x) = f(x; m, q)$ for $m = 1$ and $q = 1/4, 1/2$ and $3/4$. Similarly, Fig 2.2 shows its graphs for $q = 1/2$ and $m = 1/2, 1$ and 2.

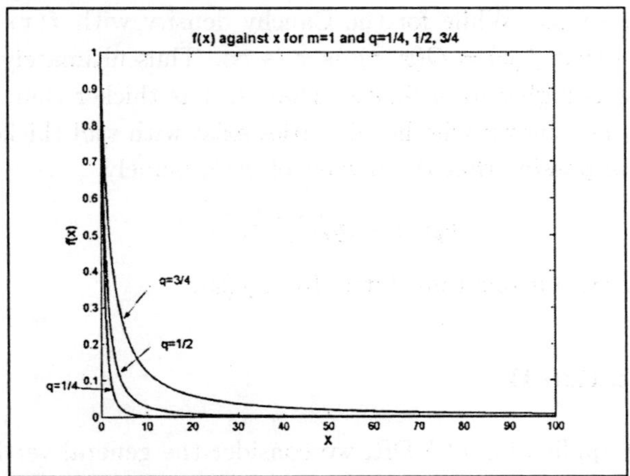

Figure 2.1: Graphs of $f(x; m, q)$ for $m = 1$ and $q = 1/4$,
1/2 and 3/4 in left to right order

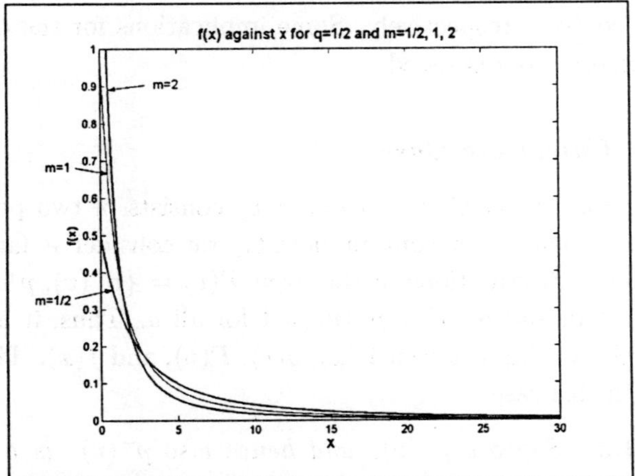

Figure 2.2: Graphs of f(x; m,q) for $q = 1/2$ and m =1/2,
1 and 2 in left to right order

For the Cauchy density, we see that

$$g(v) = \pi^{-1}[\pi v(1 - \pi v)]^{-1/2}. \tag{2.6}$$

Therefore, as $v \to 0$, $g(v) = O(v^{-\frac{1}{2}})$ and we may expect that using $g(v) = Cv^{-q}$ as above, say with, $q = \frac{3}{4}$, might lead to thicker tail behaviour than that of the Cauchy density. That is, with that choice, $g(v) = O(v^{-\frac{3}{4}})$ as $v \to 0$, which appears to be steeper than the corresponding vertical density of the Cauchy density. This is the case. The resulting $f(x)$ derived above, (2.5) is more flat-tailed than the Cauchy. That is, we have $f(x) = O(x^{-\frac{3}{4}})$ as $x \to \infty$. While for the Cauchy density with $f(x) = \pi^{-1}(1 + x^2)^{-1}$, we see that $f(x) = O(x^{-2})$ as $x \to \infty$. Thus ultimately, the tail of the new class is higher in ordinate. That is, it is thicker than that of the Cauchy. It is not known whether densities exist with still thicker tails.

We note in passing that the inverse of $g(v)$, namely,

$$p_y(y) = (y/C)^{-\frac{1}{q}}, \tag{2.7}$$

is also a density, but one that differs from $f(x)$.

2.4   Application II

For a second application of VDR, we consider the general vertical density representation concepts for univariate densities. This is called the finite contour case. The results are applied to correlated unimodal densities on $\Re$. Correlation can be decomposed into two distinct components called vertical and contour, respectively. Some implications for the consensus of correlated experts are discussed.

2.4.1   *The Univariate Case*

Consider the case in which $\{x : V(x) = v\}$ consists of two points, $x^-(v)$ and $x^+(v)$. In place of a contour density, we consider a family of discrete probability distributions of the form $P(v) = \{p^-(v), p^+(v)\}$, where $p^-(v), p^+(v) \geq 0$, and $p^-(v) + p^+(v) = 1$ for all v. Thus, it is desired to specify the relationship between $V(x)$, $g(v)$, $P(v)$, and $f(x)$. The following result holds in this case.

Theorem 2.2   *Suppose $p^-(v)$, and hence also $p^+(v)$, is a continuous function and $V(x)$ is differentiable. Then $f(x) = g(V(x))|V'(x)|P(x(v))$.*

Proof. We first consider a neighborhood of $x^-(v)$. Let Δx be a small positive increment and let $v = V(x^-)$. Then

$$P(x^-(v) \leq x \leq x^-(v) + \Delta x) \doteq f(x)\Delta x. \tag{2.8}$$

Also

$$P(x^-(v) \leq x \leq x^-(v) + \Delta x) \doteq P(v \leq V(x) \leq v + \Delta v)p^-(v) \tag{2.9}$$

by the continuity assumption on $p^-(v)$. It follows that

$$f(x)\Delta x \doteq g(v)|V'(x)|\Delta x\, p^-(v) \tag{2.10}$$

and hence in the limit as $\Delta x \to 0$

$$f(x) = g(v)|V'(x)|p^-(v). \tag{2.11}$$

The same argument applies for $p^+(v)$ and hence the theorem.

Remark. The assumption of two solutions, $x^-(v)$ and $x^+(v)$, is consistent with either concavity or convexity of $V(x)$. For example, if $V(x)$ is concave then $V(x)$ is monotone increasing on $(-\infty, x_0)$ and monotone decreasing on (x_0, ∞) for some x_0. In this case, $V'(x)$ is positive at $x^-(v)$, and negative at $x^+(v)$. Also, it is only necessary that $p^-(v)$ be continuous on $(-\infty, x_0)$; equivalently, $p^+(v)$ is continuous on (x_0, ∞). In this case, $f(x)$ may not be continuous at x_0. It may also be noted that the following corollary holds.

Corollary 2.1 *Let $V(x)$ have range $[0, \infty)$ with $v(x_0) = 0$, and suppose for each $v \geq 0$ there are two roots, $x^-(v)$ and $x^+(v)$. Then $P(x \leq x_0)$ is the weighted average of $p^-(v)$ with respect to the density, $g(v)$. Similarly, $P(x \geq x_0)$ is the same weighted average of $p^+(v)$.*

Proof.

$$\begin{aligned}
P(x \leq x_0) &= \int_{-\infty}^{0} g(V(x))|V'(x)|p^-(V(x))dx \\
&= \int_{0}^{\infty} g(v)p^-(v)dv
\end{aligned} \tag{2.12}$$

The last integral may be considered to be the weighted average of $p^-(V(x))$ on $(-\infty, x_0)$.

Remark. Since $p^+(v) = 1 - p^-(v)$ for this case, this result also verifies that $f(x)$ is a density on $(-\infty, \infty)$. The results of this section can be generalized directly to the case of more than two roots of $V(x) = v$. In this case, $P(v)$ may have a different number of components depending on the value of v.

In the next section, an application of these results is made to the decomposition of correlation into vertical and contour components.

2.5   Vertical and Contour Components of Correlation

Consider for simplicity two standard normal variables; however, the analysis can be extended, in principle, to other densities. Suppose that these variates have correlation, ρ. A standard normal variate may be given a VDR representation of the type in the previous section as follows. We have for each variate, call them X_1 and X_2, the common pdf,

$$f(x) = \frac{1}{\sqrt{2\pi}}\, e^{-\frac{x^2}{2}}. \tag{2.13}$$

Let $V(x) = |x|$. Here, $A(v) = 2v$ and $A'(v) = 2$. It follows from Theorem 1.2 that

$$g(v) = 2f(x(v)) = \sqrt{\frac{2}{\pi}}\, e^{-\frac{v^2}{2}}, \quad v \geq 0. \tag{2.14}$$

Hence, both normal variates have a vertical density representation given by $V(x) = |x|$, (2.14) and $P = \{\frac{1}{2}, \frac{1}{2}\}$. However, with this representation, there are two potential sources of dependency, one in the two identical $g(v)$-densities, and another in the two identical $P(v)$-distributions. Let the two random variables having the $g(v)$-densities be denoted V_1 and V_2. Dependency between the V_1 and V_2 random variables can be measured by a correlation, ρ_v. Dependency in the $P_1 = P_2 = P = \{\frac{1}{2}, \frac{1}{2}\}$ distributions can be measured by conditional probability. Some notations and preliminary results are needed for the derivation.

Define the following two events.

$$A \;-\; \text{the event that } \; x_1 \leq 0 \tag{2.15}$$

$$B \;-\; \text{the event that } \; x_2 \leq 0 \tag{2.16}$$

By the symmetry of the normal pdf, we have that $P(A) = P(A^c) = P(B) = P(B^c) = 1/2$. Next, define

$$\alpha = P(A|B) \tag{2.17}$$

and

$$\beta = P(A^c|B^c). \tag{2.18}$$

The joint distribution of these events may be written as

$$
\begin{array}{c c}
 & B \qquad\qquad\qquad B^c \\
\begin{array}{c} A \\ A^c \end{array}
\left[\begin{array}{c c}
\frac{1}{2}\alpha & \frac{1}{2}(1-\beta) \\
\frac{1}{2}(1-\alpha) & \frac{1}{2}\beta
\end{array} \right].
\end{array}
\tag{2.19}
$$

Since the marginal probabilities are all $1/2$, it follows that $\alpha = \beta$. Let γ be their common value.

Next, it can be easily checked that for random variables V_1 and V_2,

$$E(V_1) = E(V_2) = \sqrt{\frac{2}{\pi}}, \tag{2.20}$$

and

$$\sigma_{v_1} = \sigma_{v_2} = \sqrt{1 - 2/\pi}. \tag{2.21}$$

Define ρ_v as the correlation between V_1 and V_2. Direct calculation gives

$$E(V_1 V_2) = \frac{2}{\pi} + \rho\left(1 - \frac{2}{\pi}\right). \tag{2.22}$$

Finally, let ρ_x denote the correlation between X_1 and X_2. We next compute the correlation between X_1 and X_2, noting that their means are zero, and their standard deviations are unity.

$$
\begin{aligned}
\rho_x &= Cov(X_1, X_2) \\
&= E(X_1 X_2) \\
&= P(A \cap B)E((-V_1)(-V_2)) + P(A \cap B^c)E((-V_1)(V_2)) \\
&\quad + P(A^c \cap B)E((V_1)(-V_2)) + P(A^c \cap B^c)E(V_1 V_2) \\
&= \frac{1}{2}\gamma E(V_1 V_2) - \frac{1}{2}(1-\gamma)E(V_1 V_2) \\
&\quad - \frac{1}{2}(1-\gamma)E(V_1 V_2) + \frac{1}{2}\gamma E(V_1 V_2)
\end{aligned}
$$

$$= (2\gamma - 1)E(V_1 V_2). \tag{2.23}$$

Thus noting (2.22), we have finally that

$$\rho_x = (2\gamma - 1)(2/\pi + \rho_v(1 - 2/\pi)). \tag{2.24}$$

Some interesting implications of this result can be readily observed. For $\gamma = 1/2$, the contour distributions, P_1 and P_2 are independent and this case is analogous to zero correlation. Also, in this case, the x-random variables are independent ($\rho_x = 0$) irrespective of ρ_v. If $\rho_v = +1$, and hence the V-random variables are fully correlated, then the correlation of the X-random variables is identical to that of the contour random variables. It seems sensible in both these cases to define the *contour correlation* by ρ_c where

$$\rho_c = 2\gamma - 1 \tag{2.25}$$

so that (2.24) may be written as

$$\rho_x = \rho_c(2/\pi + \rho_v(1 - 2/\pi)) \ . \tag{2.26}$$

Since $0 \le \gamma \le 1$, we have $-1 \le \rho_c \le 1$.

Although the name, contour correlation, is expressive of the derivation, it may be noted that ρ_c could also be called either the *sign correlation* or the *error direction correlation*. Table 2.1 gives some representative values.

Table 2.1: Values of ρ_x corresponding to those of ρ_c and ρ_v

ρ_v ρ_c	-1.0	-0.5	0.0	+0.5	+1.0
-1.0	-0.2732	-0.4549	-0.6366	-0.8183	-1.0
-0.5	-0.1366	-0.2275	-0.3183	-0.4092	-0.5
0.0	0.0	0.0	0.0	0.0	0.0
+0.5	0.1366	0.2275	0.3183	0.4092	0.5
+1.0	0.2732	0.4549	0.6366	0.8183	1.0

It is interesting to note that the sign of the correlation ρ_x is completely determined by that of ρ_c, since the minimum value of the term in parentheses is $4/\pi - 1 > 0$. Thus, two such variates could have zero correlation due to $\rho_c = 0$ and yet have any possible value of ρ_v in $[-1, 1]$. Some implications and other examples are discussed further below in the context of expert estimation error distributions.

2.6 Discussion

Although developed above in terms of standard normal variables, the same constructs can be similarly applied, in principle, to any univariate densities. The derivation from (2.20) - (2.25) would need to be customized to the particular V_1 and V_2 densities under consideration. Specifically, their means and variances would need to be computed as well as $E(V_1 V_2)$. If the densities were specified initially in $f(x)$ form, i.e. in the observation space, then Theorem 2.1 would need to be applied first to obtain the corresponding $g(v)$-pdfs. However, these steps should not be particularly difficult. In this more general setting, ρ_c measures the extent to which two random variables tend to simultaneously be above or below their mean values; while ρ_v measures the extent to which the errors are large or small together. These results may be especially useful in the aggregation or reconciliation of expert estimates.

The Bayesian theory of expert aggregation has been developed by Winkler (1968, 1986), Clemen and Winkler (1985, 1990, 1993), and others - see also Ganest and Zidek (1986) and West (1988). Correlation of expert errors plays a central role in these Bayesian approaches. The present results provide a direction for more detailed analysis in this setting by permitting expert errors to have more than one kind of correlation at the same time.

The Bayesian theory generally tends to decrease the effective weights of correlated experts in the final aggregated estimates. This is sensible on intuitive grounds since high correlation suggests that no, or less, new information is gained by including one or the other such expert. Put differently, inclusion of both correlated experts on an equal weights basis would, in effect, double-weight just one independent estimate. Therefore we expect aggregation methods to react by decreasing the assigned weights. At the other extreme of zero correlation between two expert error densities, they are treated equally and tend to receive about equal weights in the final aggregate. This also accords well with intuition. However, there are five cases, depending on the values of ρ_x and ρ_v, which raise doubts about these results. In what follows, we assume a focus on two experts out of a possible larger sample; and all correlations, ρ_c, ρ_v, and ρ_x, are zero between each of the focus pair and each of the others.

Case 1: $\rho_x = 0, \rho_v = -1$. From Table 2.2, the correlation of $\rho_x = 0$ essentially hides or can be associated with any possible value of ρ_v. Consider the extreme case of $\rho_v = -1$. Since $\rho_x = 0$, variables X_1 and X_2 are independent. Equal weightings of all the experts would be called for both by the Bayesian approach and by intuition. But since $\rho_v = -1$, we have that V_1 and V_2 are negatively correlated. If expert one is very accurate (large v-score) then expert two is very inaccurate (small v-score) and vice versa. Intuition suggests in this case that only one or the other should be retained. An obvious selection criterion can be based on *imputed error*. Namely, suppose expert one is retained first and expert two is omitted. An equal weighting of the estimates x_i of the retained set yields the potential aggregate w_1, say. Similarly, if expert two is retained, then an aggregate w_2 is obtained. The smaller of the two errors, $e_1 = (x_1 - w_1)^2$ and $e_2^2 = (x_2 - w_2)^2$, provides a basis for a choice. As an example, let four expert estimates, in expert number order, be $\{1, 2, 5, 10\}$. Then $w_1 = 16/3, w_2 = 17/3, e_1 = (13/3)^2$, and $e_2 = (11/3)^2$. Thus, this choice is in favor of expert two with $w = w_2 = 17/3$. That is, expert two is chosen, expert one is discarded, and the estimate is $w_2 = 17/3$.

Case 2: $\rho_x = 0, \rho_v = 1$. Here $\rho_x = 0$ is evidently due to $\rho_c = 0$ from table 2.2. But the perfect correlation between V_1 and V_2 shows that the errors of expert one and expert two must always be identical in magnitude. It follows that the only aggregate w consistent with the data set $\{1, 2, 5, 10\}$ is 1.5, the mean of expert one and expert two's estimates. Evidently, such information would be extremely strong in its effect on aggregation.

The next three cases consider ρ_x (and also ρ_c) to have maximum value for a given value of ρ_v.

Case 3: $\rho_x = $ maximum $= 0.2732, \rho_v = -1$. From Table 2.2, we see that this implies $\rho_c = 1$. This case is representative of very large values of ρ_c

with very small values of ρ_v. In this particular extreme case, the $\rho_v = -1$ considerations of Case 1 apply again.

Case 4: $\rho_x = $ maximum $ = 0.6366, \rho_v = 0$. Here, $\rho_c = 1$ as in Case 3. This case is an extreme version of very high error direction correlation along with zero accuracy correlation ρ_v. The large ρ_c and moderately positive overall ρ_x-correlation argue for discounting one or both of these experts. However, since $\rho_v = 0$, neither error size gives any information about the other. Therefore, all that can be said at the present level of analysis is that whatever be the errors imputed to these experts, the signs of such errors must be identical.

Case 5: $\rho_x = 1.0, \rho_v = 1.0$. It follows that $\rho_c = 1$. This case is the extreme version with cases of high positive correlations of all three types. When $\rho_v = 1.0$, the same considerations apply as in Cases 2 and 3. However, with ρ_v merely near to 1.0, the two expert errors no longer need to be identical. They need only to be near to each other with high probability. However, whatever error values are imputed to them, their signs should also be the same with high probability.

The analysis of the foregoing cases shows that for one simple situation in which a pair of experts can be isolated from the rest of a sample, the consensus point estimate depends not only on the ordinary correlation or ρ_x, but also is particularly sensitive to the ρ_v-correlation. This is also to be expected from another viewpoint. In the Bayesian approach, accuracy, with respect to bias, is handled by initial adjustment of error distributions for known biases if any. Accuracy, as precision, is measured only by variances. Collection and use of $\boldsymbol{\rho_v}$ provides an important additional source of information on precision. In fact, for Case 2 above, the $\boldsymbol{\rho_v}$-information was sufficient, along with other assumptions to actually identify an appropriate consensus point.

These results indicate the need for a refinement of the Bayesian approach to reflect the impacts of the $\boldsymbol{\rho_v}$ and $\boldsymbol{\rho_c}$ correlation matrices. Alternatively, an approach that directly assigns appropriate weights might be sought, in which the weights attempt to reflect the implications of the $\boldsymbol{\rho_v}$ and $\boldsymbol{\rho_c}$ correlation matrices. A further important consideration, signalled particularly by Cases 2 and 3 and to a lesser extent by Cases 4 and 5, is that weights for aggregating individual experts into a consensus point estimate should apparently depend on the precision assigned to that expert by the weighting method.

A simple operationalization of such dependence of weights on imputed accuracies can be given as follows without reference to correlations. Let x_c be the desired consensus point; and suppose that $x_c = \sum w_i x_i$ where $w_i \geq 0$ and $\sum w_i = 1$. The error imputed to expert-i if x_i is chosen as the consensus can be measured by $(x_c - x_i)^2$. Then a measure of the imputed precision or accuracy of expert i can be taken as $\exp\{-(x_c - x_i)^2\}$. If weights w_i are required to be proportionate to these precision measures, then the condition required on x_c can be seen to be

$$x_c \sum_i \exp\{-(x_c - x_i)^2\} = \sum x_i \exp\{-(x_c - x_i)^2\} \qquad (2.27)$$

which can be solved by the Newton-Raphson method with, for example, $\overline{x}$ as a starting value.

2.7   Further Considerations

This chapter discusses vertical density representation on $\Re$ and some consequences for correlated normal variables. Results for the general univariate case are derived here. These results were applied to the analysis of correlation. Correlation can be analyzed into contour and vertical components, thus permitting more detailed analyses involving correlated variables.

These results also give some insight on the well known difficulty of defining the multivariate gamma density. See for example, Law and Kelton (1982). Let ρ^c denote the matrix of contour correlations, ρ^v the matrix of vertical correlations, and ρ^x denote the overall correlation matrix. Evidently for one and the same multivariate normal distribution with correlation matrix ρ^x, one could associate a continuum of multivariate gamma densities as ρ^c and ρ^v vary in such a way that their combined effect is ρ^x.

Specifically, let X be the bivariate normal, $N(0, \Sigma)$, with correlation matrix, ρ^x. Let $V_1 = X_1^2$ and $V_2 = X_2^2$ with correlation matrix, ρ^v. Let ρ^c be the contour correlation matrix for X_1 and X_2. Then any number of such bivariate gamma distributions can be associated with the $N(0, \Sigma)$ density provided ρ^c is chosen so that ρ^c and ρ^v yield ρ^x.

The decomposition of correlation into separate components for accuracy and error direction, respectively, promises to be especially useful for the

Bayesian consensus of experts. In particular, these considerations may be helpful in resolving the intuitive dilemma arising when highly accurate experts are discounted due to correlation.

Chapter 3

Multivariate Vertical Density Representation

Introduction

In this chapter, we develop the theory of vertical density representation (VDR) in the multivariate case. We present a formula for the calculation of the conditional probability density of a random vector given its density value. Most of the materials given here are based on Pang *et al.* (2001), Kotz *et al.* (1997), and Kozubowski (2002).

As we have seen, the concept of VDR as it was originally developed, was closely related to the generation of normal random variates. Let X_1 and X_2 be two independent standard normal distributions on $\Re$; then for given $r = \sqrt{x_1^2 + x_2^2}$, the conditional distribution of $\mathbf{X}_2 = (X_1, X_2)$ is the uniform distribution on the circle $\{\mathbf{x}_2 = (x_1, x_2) : x_1^2 + x_2^2 = r^2\}$. The density of $\mathbf{X}_2 = (X_1, X_2)$ is

$$f(\mathbf{x}_2) = \frac{1}{2\pi} \exp\left(-\frac{1}{2}\left[x_1^2 + x_2^2\right]\right).$$

It was shown in Section 1.5 of Chapter 1 that the density of $R = \sqrt{-2\ln(2\pi V)}$ is $f_R(r) = r\exp\left(-\frac{r^2}{2}\right)$, while the conditional density of $\mathbf{X}_2 = (X_1, X_2)$ given r is

$$h_R(\mathbf{x}_2 \mid r) = \frac{1}{2\pi r}\delta_r\left(\sqrt{x_1^2 + x_2^2}\right). \tag{3.1}$$

45

Therefore the density of $\mathbf{X}_2$ can be represented as,

$$f(\mathbf{x}_2) = \int_0^\infty h_R(\mathbf{x}_2 \mid r) f_R(r) dr,$$

or equivalently

$$f(\mathbf{x}_2) = \int_0^m h_V(\mathbf{x}_2 \mid v) g(v) dv, \tag{3.2}$$

where $m = \sup\left\{ f(\mathbf{x}_2) : \mathbf{x}_2 \in \Re^2 \right\}$, $h_V(\cdot \mid v)$ is the conditional pdf of $\mathbf{X}_2$ given $V = v$, and $g(\cdot)$ is the density of $V = f(\mathbf{X}_2)$. Troutt and Pang (1997) obtained a smooth solution $h_V(\cdot \mid v)$ for equation (3.2) in the one dimensional case. It is natural to ask whether we can generalize the above procedure to the d -dimensional case with $d \geq 3$? As we shall see, the answer is positive.

It is perfectly possible to give a representation of a density function, $f(\mathbf{x}_d)$, of a d-dimensional normal vector $\mathbf{X}_d$ in a form similar to (3.2) such that the conditional distribution of the vector $\mathbf{X}_d$ given density value $f = v$ is uniform on the surface of a d-dimensional sphere with radius $r = r(v)$. More generally, we can find the conditional distribution given density value $f = v$ and it is also possible to generate a smooth density pair (h, g) as solutions of equation (3.2) for the case of the k-dimensional normal density by taking certain k components of a d-dimensional normal vector with $d > k$. The above findings are valid for a very wide family of distributions.

3.1   Multivariate VDR and Conditional Density

In this section we consider two types of VDR. Assume that the d-dimensional random vector $\mathbf{X}_d$ has pdf $f(\mathbf{x}_d)$ and let $V = f(\mathbf{X}_d)$. Type I VDR was proposed by Troutt (1991). Let L_d be the Lebesgue measure in $\Re^d$, and $S_f(v) = \{\mathbf{x}_d : f(\mathbf{x}_d) \geq v\}$. If $L_d(S_f(v))$ is differentiable, then the density of V is

$$g(v) = -v \frac{d}{dv} L_d(S_f(v)).$$

Moreover, the density $f(\mathbf{x}_d)$ can be expressed in the form

$$f(\mathbf{x}_d) = \int\limits_0^{f_0} h_V(\mathbf{x}_d \mid v)g(v)dv,$$

where $h_V(\mathbf{x}_d \mid v)$ is the conditional density of $\mathbf{X}_d$ given $f = v$, and $f_0 = \sup\{f(\mathbf{x}_d) : \mathbf{x}_d \in \Re^d\}$.

Type II VDR was given by Fang *et al.* (2001). It is best summarized by the following.

Lemma 3.1 *Let $D_{d+1}(f)$ be the set in $\Re^{d+1}$ defined by*

$$D_{d+1}(f) = \{\mathbf{x}_{d+1} = (\mathbf{x}_d, x_{d+1}) : 0 \le x_{d+1} \le f(\mathbf{x}_d)\}$$

and let $\mathbf{X}_{d+1} = (\mathbf{X}_d, X_{d+1})$ have the uniform distribution on $D_{d+1}(f)$. Then
(i) The pdf of $\mathbf{X}_d$ is $f(\mathbf{x}_d)$;
(ii) The density of X_{d+1} is $L_d(S_f(x))$;
(iii) The conditional distribution of $\mathbf{X}_d$ given $X_{d+1} = v$ is the uniform distribution on $S_f(v)$.

In the sequel, we proceed to find $h_V(\mathbf{x}_d \mid v)$ using the Type II VDR. We define

$$\Gamma_f(v) = \{\mathbf{x}_d : f(\mathbf{x}_d) = v\}, \tag{3.3}$$

which is a level surface of f and we have the following theorem (Pang *et al.*, 2001).

Theorem 3.1 *Let $f(\mathbf{x}_d)$ be the pdf of a random vector $\mathbf{X}_d$. Assume that*
(i) $\nabla f(\mathbf{x}_d)$ is continuous and nonzero on $\Gamma_f(v)$; and
(ii) for any fixed unit vector $\mathbf{d} \in \Re^d$, $f(r\mathbf{d})$ is a decreasing function in r. Then

$$p(\mathbf{x}_d \mid \mathbf{x}_d \in \Gamma_f(v)) = \frac{I_{\Gamma_f(v)}}{\|\nabla f(\mathbf{x}_d)\|} \left(\int_{\Gamma_f(v)} \frac{1}{\|\nabla f(\mathbf{x}_d)\|} ds \right)^{-1}$$

where $\|\nabla f(\mathbf{x}_d)\| = \sqrt{\sum_{i=1}^n \left[\frac{\partial f}{\partial x_i}(\mathbf{x}_d) \right]^2}$, and $d\mathbf{s}$ is the corresponding volume form on the set $\Gamma_f(v)$.

Remark. It is well known that on the $(d-1)$-dimensional surface $\Gamma_f(v)$, the volume form is

$$\mathbf{ds} = \frac{\|\nabla f\|}{\left|\frac{\partial f}{\partial x_d}\right|} dx_1 dx_2 \ldots dx_{d-1},$$

where x_d is considered to be a function of $\mathbf{x}_{d-1} = (x_1, \ldots, x_{d-1})$, being defined implicitly by $f(\mathbf{x}_{d-1}, x_d) = v$.

Corollary 3.1   *Let $f(\mathbf{x}_d)$ be a spherically symmetric pdf in the form $f(\mathbf{x}_d)$ $= w(\mathbf{x}_d\mathbf{x}_d')$, where $w(\cdot)$ is a strictly decreasing and differentiable real function. If $\mathbf{X}_d$ has pdf $f(\mathbf{x}_d)$, then the conditional pdf of $\mathbf{X}_d$ given $f = v$ is uniformly distributed on $\{\mathbf{x}_d : f(\mathbf{x}_d) = v\}$.*

The Corollary follows from Theorem 3.1 by noting that

$$\|\nabla f(\mathbf{x}_d)\|^2 = (w'(w^{-1}(v)))^2 \sum_{i=1}^{d} 4x_i^2 = 4(w'(w^{-1}(v)))^2(w^{-1}(v)),$$

which is constant on $\Gamma_f(v)$.

Corollary 3.2   *The conditional distribution $\mathbf{X}_d$ given $f(\mathbf{X}_d) = v$ is uniform on the surface $f(\mathbf{x}_d) = v$ if and only if $\|\nabla f\| = u(f)$, where u is a continuous function.*

Example 3.1. Let the random vector $\mathbf{X}_2 = (X_1, X_2)$ have the pdf

$$f(x_1, x_2) = \frac{1}{2\pi\sigma_1\sigma_2} \exp\left(-\frac{x_1^2}{2\sigma_1^2} - \frac{x_2^2}{2\sigma_2^2}\right),$$

with $\sigma_1, \sigma_2 > 0$ and $\sigma_1 \neq \sigma_2$. The curve $\Gamma_f(v)$ is defined by the equation

$$f(x_1, x_2) = \frac{1}{2\pi\sigma_1\sigma_2} \exp\left(-\frac{x_1^2}{2\sigma_1^2} - \frac{x_2^2}{2\sigma_2^2}\right) = v. \tag{3.4}$$

Since

$$\mathbf{ds} = \frac{\|\nabla f\|}{\left|\frac{\partial f}{\partial x_2}\right|} dx_1 \quad \text{on } \Gamma_f(v),$$

we have

$$\int_{\Gamma_f(v)} \frac{\mathbf{ds}}{\|\nabla f\|} = \int_{f=v} \frac{dx_1}{\left|\frac{\partial f}{\partial x_2}\right|}$$

$$= \int_{f=v} \frac{\sigma_2^2 dx_1}{v\sqrt{x_2^2}}$$

$$= \int_{-\sqrt{2}a\sigma_1}^{\sqrt{2}a\sigma_1} \frac{\sigma_2 dx_1}{\sqrt{2}va\sqrt{1 - \left(\frac{x_1}{\sqrt{2}a\sigma_1}\right)^2}}$$

$$= \frac{\pi\sigma_1\sigma_2}{v},$$

where $a = \sqrt{-\ln(2\pi\sigma_1\sigma_2 v)}$. Hence,

$$p\left((x_1, x_2) \mid f = v\right) = \frac{\frac{1}{\|\nabla f\|}}{\int_{\Gamma_f(v)} \frac{ds}{\|\nabla f\|}} I_{\{f=v\}}$$

$$= \left(\sqrt{\frac{v^2 x_1^2}{\sigma_1^4} + \frac{v^2 x_2^2}{\sigma_2^4}}\right)^{-1} \bigg/ \frac{\pi\sigma_1\sigma_2}{v} I_{\{f=v\}}$$

$$= \frac{1}{\pi\sigma_1\sigma_2\sqrt{\frac{x_1^2}{\sigma_1^4} + \frac{x_2^2}{\sigma_2^4}}} I_{\{f=v\}}$$

3.2   Some Results on the Multivariate Normal Distribution

The conditional density, $h_V(\mathbf{x}_d \mid v)$, in Theorem 3.1 is clearly degenerate. It only assumes nonzero values on a set of measure zero in $\Re^d$. Is there a non-degenerate pair, $(h_V(\mathbf{x}_d \mid v), g(v))$, such that

$$f(\mathbf{x}_d) = \int_0^{f_0} h_V(\mathbf{x}_d \mid v)g(v)dv,$$

where $f_0 := \sup\{f(\mathbf{x}_d) : \mathbf{x}_d \in \Re^d\}$? We now give a non-degenerate pair for the case of the multivariate normal distribution.

Let the random vector $\mathbf{X}_d$ have a multivariate normal distribution with density function

$$f(\mathbf{x}_d) = \frac{1}{(2\pi)^{\frac{d}{2}}} \exp\left(-\frac{1}{2}\mathbf{x}_d\mathbf{x}_d'\right) = \frac{1}{(2\pi)^{\frac{d}{2}}} \exp\left(-\frac{1}{2}\sum_{j=1}^d x_j^2\right).$$

Kotz *et al.* (1997) have given the probability density function of $V = f(\mathbf{X}_d)$ as

$$g(v) = \frac{2\pi^{d/2}}{\Gamma(d/2)}[-2\ln((2\pi)^{d/2}v]^{\frac{d}{2}-1}, \qquad 0 \le v \le f_0 = \left(\frac{1}{2\pi}\right)^{\frac{d}{2}}.$$

By Corollary 3.1, $h_V(\mathbf{x}_d \mid v)$ is uniformly distributed on the sphere with radius $r = \sqrt{-2\ln\left((2\pi)^{\frac{d}{2}}v\right)}$. Let the random vector $\mathbf{Z}_d = (Z_1, Z_2, \ldots, Z_d)$ have the uniform distribution on the unit sphere $\sum_{i=1}^{d} Z_i^2 = 1$. Then $r\mathbf{Z}_d$ has the uniform distribution on the spherical surface

$$\bar{S}_d(r) = \left\{ \mathbf{x}_d \in \Re^d : \sum_{i=1}^{d} x_i^2 = r^2 \right\}.$$

Thus we have the well known stochastic representation of the multivariate normal distribution

$$\mathbf{X}_d = \sqrt{-2\ln\left((2\pi)^{\frac{d}{2}}V\right)}\,\mathbf{Z}_d, \quad V \sim g, \quad \mathbf{Z}_d \sim U(\bar{S}_d(1)).$$

$$\mathbf{X}_d = \sqrt{-2\ln\left((2\pi)^{\frac{d}{2}}V\right)}\,\mathbf{Z}_d, \quad V \sim g, \quad \mathbf{Z}_d \sim U(\bar{S}_d(1)).$$

Let k be an integer such that $k < d$ and for the first k components of $\mathbf{X}_d$

$$\mathbf{X}_k = (X_1, \ldots, X_k) = \sqrt{-2\ln\left((2\pi)^{\frac{d}{2}}V\right)}(Z_1, \ldots, Z_k) = \sqrt{-2\ln\left((2\pi)^{\frac{d}{2}}V\right)}\,\mathbf{Z}_k.$$

The pdf of $\mathbf{X}_k$ is

$$f_k(\mathbf{x}_k) = \int\limits_{\{r(v) \ge \|\mathbf{x}_k\|\}} \frac{1}{r^k(v)} b_{d,k}\left(\frac{\mathbf{x}_k}{r(v)}\right) g_d(v)\,dv, \qquad (3.5)$$

where $r(v) = \sqrt{-2\ln\left((2\pi)^{\frac{d}{2}}v\right)}$, $\quad 0 \le v \le f_0 = (1/2\pi)^{\frac{d}{2}}$ and $b_{d,k}$ is the marginal probability density of the first k components of random vector $\mathbf{Z}_d$ that has the uniform distribution on the unit sphere in $\Re^d$.

Let $\mathbf{Z}_k$ be the first k components of the random vector $\mathbf{Z}_d$. Its marginal pdf is

$$
\begin{aligned}
b_{d,k}(\mathbf{z}_k) \;=\;& \frac{1}{L_d(\bar{S}_d(1))} \int_D \frac{1}{\sqrt{1 - \sum_{i=1}^{d-1} z_i^2}} \, dz_{k+1} \cdots dz_{d-1} \\[2mm]
=\;& \frac{1}{L_d(\bar{S}_d(1))} \frac{1}{\sqrt{1 - \sum_{i=1}^{k-1} z_i^2}} \int_D \frac{\sqrt{1 - \sum_{i=1}^{k-1} z_i^2}}{\sqrt{1 - \sum_{i=1}^{d-1} z_i^2}} \, dz_{k+1} \cdots dz_{d-1} \\[2mm]
=\;& \frac{1}{L_d(\bar{S}_d(1))} \frac{1}{\sqrt{1 - \sum_{i=1}^{k-1} z_i^2}} L_{d-k}\left(\bar{S}_{d-k}\left(\sqrt{1 - \sum_{i=1}^{k-1} z_i^2} \right) \right) \\[2mm]
=\;& \frac{\Gamma\left(\frac{d}{2}\right)}{\pi^{k/2} \Gamma\left(\frac{d-k}{2}\right)} \left[\sqrt{1 - \sum_{i=1}^{k} z_i^2} \right]^{d-k-2}, \\[2mm]
=\;& \frac{\Gamma\left(\frac{d}{2}\right) 2\pi^{(d-k)/2}}{2\pi^{d/2} \Gamma\left(\frac{d-k}{2}\right)} \left[\sqrt{1 - \sum_{i=1}^{k} z_i^2} \right]^{d-k-2}
\end{aligned}
$$

where $D = \left\{ (Z_{k+1}, \ldots, Z_n) : \sum_{i=k+1}^{d} Z_i^2 = 1 - \sum_{i=1}^{k} Z_i^2 \right\}$ and $L_m(\bar{S}_m(r))$ denotes the surface area of the sphere $\bar{S}_m(r)$ in $\Re^m$ with radius r. Substituting $b_{d,k}$ into (3.5) gives

$$
\begin{aligned}
f_k(\mathbf{x}_k) \;=\;& \int_0^{v^*} \frac{2\pi^{\frac{d-k}{2}-1}}{\Gamma\left(\frac{d-k}{2}\right)} \left[-2\ln((2\pi)^{d/2} v) \right]^{\frac{d}{2}-1} \frac{1}{(-2\ln((2\pi)^{d/2} v))^{\frac{k}{2}}} \\[2mm]
& \left(1 - \frac{\mathbf{x}_k \mathbf{x}_k'}{-2\ln((2\pi)^{d/2} v)} \right)^{\frac{d-k}{2} - \frac{1}{2}} dv \\[2mm]
=\;& \int_0^{v^*} \frac{2\pi^{\frac{d-k}{2}}}{\Gamma\left(\frac{d-k}{2}\right)} \left[-2\ln((2\pi)^{d/2} v) \right]^{\frac{d-k}{2}-1} \cdot \\[2mm]
& \left(1 - \frac{\mathbf{x}_k \mathbf{x}_k'}{-2\ln((2\pi)^{d/2} v)} \right)^{\frac{d-k}{2} - \frac{1}{2}} dv, \qquad\qquad (3.6)
\end{aligned}
$$

where v^* is defined by $-2\ln\left((2\pi)^{\frac{d}{2}} v^* \right) = \mathbf{x}_k \mathbf{x}_k'$. It is not difficult to check that the above integral (3.6) reduces to the k-dimensional normal density:

$$\int_0^{s^*} \frac{1}{2^{\frac{d}{2}-1}\Gamma\left(\frac{d-k}{2}\right)(\pi)^{\frac{k}{2}}}(-2\ln s)^{\frac{d-k}{2}-1}\left(1-\frac{\mathbf{x}_k\mathbf{x}_k'}{-2\ln s}\right)^{\frac{d-k}{2}-1}ds \quad (s=(2\pi)^{d/2}v)$$

$$=\int_0^{s^*}\frac{1}{2^{\frac{d}{2}-1}\Gamma\left(\frac{d-k}{2}\right)(\pi)^{\frac{k}{2}}}(-2\ln s-\mathbf{x}_k\mathbf{x}_k')^{\frac{d-k}{2}-1}ds \quad (t=-2\ln s-\mathbf{x}_k\mathbf{x}_k')$$

$$=\exp\left(-\frac{\mathbf{x}_k\mathbf{x}_k'}{2}\right)\int_0^{\infty}\frac{1}{2^{\frac{d}{2}-1}\Gamma\left(\frac{d-k}{2}\right)\pi^{\frac{k}{2}}}t^{\frac{d-k}{2}-1}e^{-\frac{1}{2}}\frac{dt}{2}=\frac{1}{(2\pi)^{\frac{k}{2}}}\exp\left(-\frac{\mathbf{x}_k\mathbf{x}_k'}{2}\right),$$

where $s^*=(2\pi)^{\frac{d}{2}}v^*$.

When $k=1$, we recover the representation of the standard normal distribution function. So this result is an extension of Troutt and Pang (1997).

3.3   An Application of Multivariate VDR

In this section, we use the results of Theorem 3.1 to devise a method to generate random vectors, which follows a certain multivariate distribution.

The purpose of uniform random vector generation is to simulate a sequence of independent random vector variables with the uniform distribution on $\mathbf{I}^d=[0,1]^d$, $d\geq 2$, as their common multivariate distribution function. Anderson (1990), Bhavsar and Issac (1987) and Eddy (1990) and among others, have presented articles on uniform random vector generation for parallelized simulation methods. Also, Eddy (1994) gave an excellent discussion for the background on issues related to random vector generation and parallelized simulation. The three most common methods to generate uniform random vectors are (i) the matrix method, (ii) the multiple-recursive matrix method and (iii) the inversion method. The matrix method is an analog of the classical linear congruential method ($d=1$). For the multiple-recursive matrix method and the inversion method, one may refer to the two articles by Niederreiter (1994, 1995).

In the following, we present a method for generating uniform random vectors, which is a consequence of Theorem 3.1. Let $f(\mathbf{x}_d)$ be a probability density function on $\Re^d$ and suppose that it is sufficiently smooth. The

VDR method is a general method to generate random vectors $\mathbf{X}_d$ with density f. The procedure may be described as follows:

(1) Generate random variable, V, which follows the distribution of $g(v)$ and let v be the value that V assumes;
(2) Generate vector, $\mathbf{X}_d$, which follows the distribution on the surface $\{\mathbf{x}_d \in \Re^d : f(\mathbf{x}_d) = v\}$ as described in Theorem 3.1.
(3) Deliver $\mathbf{X}_d$, whose distribution follows f.

Now we give a simple example to illustrate the above method.

Example 3.2. Let $\mathbf{X}_2 = (X_1, X_2)$ have pdf $f(\mathbf{x}_2) = h\left(|x_1|^2 + |x_2|\right)$, where h is a strictly decreasing continuous function. Clearly,

$$\Gamma_f(v) = \left\{\mathbf{x}_2 : |x_1|^2 + |x_2| = h^{-1}(v)\right\},$$

where h^{-1} is the inverse function of h. Let $a(v) = h^{-1}(v)$ and

$$S_f(v) = \left\{\mathbf{x}_2 : |x_1|^2 + |x_2| \le a(v)\right\}.$$

Firstly, we find the distribution of $f(\mathbf{X}_2)$. Define

$$S_f^{++}(v) = \{\mathbf{x}_2 : |x_1|^2 + |x_2| \le a(v), x_1 \ge 0, x_2 \ge 0\},$$

and $S_f^{+-}, S_f^{--}, S_f^{-+}$ similarly. It is easy to see that

$$S_f(v) = S_f^{++}(v) \cup S_f^{+-}(v) \cup S_f^{--}(v) \cup S_f^{-+}(v).$$

and

$$L_2(S_f^{++}) = L_2(S_f^{+-}) = L_2(S_f^{--}) = L_2(S_f^{-+})$$

so that

$$L_2(S_f(v)) = 4L_2(S_f^{++}) = 4\int_0^{\sqrt{a}} (a - x^2)dx = \frac{8}{3}(a(v))^{\frac{3}{2}},$$

and

$$g(v) = -v\frac{d}{dv}\left(\frac{8}{3}v^{\frac{3}{2}}\right) = -\frac{4v\sqrt{v}}{h'(h^{-1}(v))}. \tag{3.7}$$

$\Gamma_f(v)$ is a closed curve consisting of four segments as shown in Figure 3.1 below.

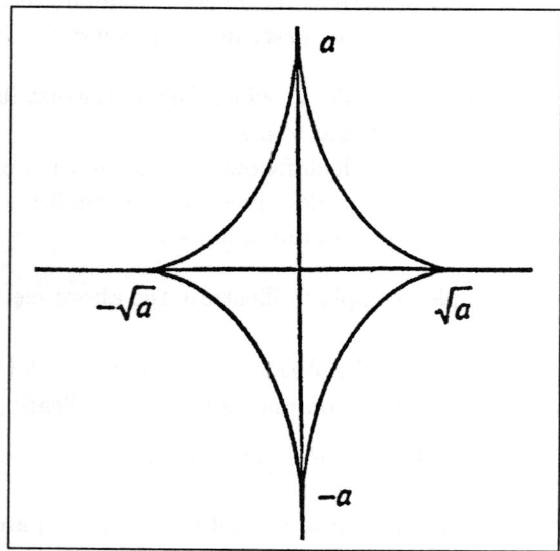

Figure 3.1: Plot of the $\Gamma_f(v)$ curve.

Let the point $\mathbf{x}_2^0 = (0, a)$, be the starting point of the curve with the first segment from $(0, a)$ to $(\sqrt{a}, 0)$, the second segment from $(\sqrt{a}, 0)$ to $(0, -a)$, the third from $(0, -a)$ to $(-\sqrt{a}, 0)$ and the fourth from $(-\sqrt{a}, 0)$ to $(0, a)$. Denote the length of the curve from $\mathbf{x}_2^0$ to $\mathbf{x}_2 = (x_1, x_2)$ by $s(\mathbf{x}_2) = s(x_1)$ given $f(\mathbf{x}_2) = v$. By taking the length of the curve as the parameter, the curve $\Gamma_f(v)$, can be written as $\mathbf{x}_2 = \mathbf{x}_2(s)$, where $s = s(x_1)$. A random vector, $\mathbf{x}_2 \in \Gamma_f(v)$, is equivalent to a random variable, T, taking value, $s = s(x_1)$, and the density of, T, is given by Theorem 3.1; namely,

$$p(s \mid f(\mathbf{x}_2) = v) = \frac{\frac{1}{\|\nabla f(\mathbf{x}_2(s))\|}}{\int_{\Gamma_f(v)} \frac{1}{\|\nabla f(\mathbf{x}_2(s))\|} ds}, \quad 0 \le s \le \sqrt{a(v)}.$$

Noticing $s = s(x_1)$ and $ds = \|\nabla f(\mathbf{x}_2(s))\| \, dx_1$ in this case, we have

$$p_1(x_1 \mid f(\mathbf{x}_2) = v) = \frac{1}{\int_o^{\sqrt{a(v)}} dt} = \frac{1}{\sqrt{a(v)}},$$

which is uniform. So we have the following algorithm for generating the random vector, which follows f.

(1) Let V be the random variable with pdf $g(\cdot)$ as given in 3.7;
(2) Let X_1 have the uniform distribution on $[0, a(V)]$ and $X_2 = a(V) - X_1^2$;
(3) Let U have the uniform distribution on $[0, 1]$. Deliver

$$\mathbf{X}_2 = \left((-1)^J X_1, (-1)^I X_2 \right),$$

where

$$J = \operatorname{sgn}\left[\frac{1}{2} - U \right] = \left\{ \begin{array}{rll} 1, & \text{if} & \frac{1}{2} - U > 0 \\ 0, & \text{if} & \frac{1}{2} - U = 0 \\ -1, & \text{if} & \frac{1}{2} - U < 0 \end{array} \right.$$

$$V = \left\{ \begin{array}{ll} 2U, & \text{if} \quad U < \frac{1}{2} \\ 2(U - \frac{1}{2}), & \text{if} \quad U > \frac{1}{2} \end{array} \right.$$

and $I = \operatorname{sgn}[\frac{1}{2} - V]$. Then $\mathbf{X}_2$ follows f.

3.4 Tail Behaviour and Multivariate VDR

In Chapter 2, we have presented the properties of tail behaviour for some common univariate distributions using the VDR method. For multivariate distributions, there exist other ways to define the tail behaviour in a proper manner mathematically. In fact, we can extend the idea of Kotz and Troutt (1996) to the multivariate distribution case. Kotz, Fang and Liang (1997) studied some random variables related to the vertical probability density functions for some basic multivariate distributions and considered generation of the distributions.

3.4.1 *Spherically Symmetric Distributions*

The spherically symmetric distributions are well-known generalizations of the multivariate normal distribution. The multivariate normal distribution can be viewed as a spherically (or elliptically) symmetric distribution. Chmielewski (1981) gave an excellent review on elliptically symmetric distributions. Earlier papers on this topic can be traced back to Bartlett

(1934), and Hartman and Winter (1940). The recent monograph by Fang, Kotz and Ng (1990) also gives a very comprehensive treatise on symmetric multivariate distributions. In Fang, Kotz and Ng (1990), another way of defining spherically symmetric distributions can be found. To begin with, we assume that the random vector $\mathbf{x} = (x_1, \ldots, x_n)'$ follows a spherically symmetric distribution with the density function $f(\mathbf{x}) = h(\mathbf{x}'\mathbf{x})$, $\mathbf{x} \in \Re^n$, where $h(\cdot)$ is strictly decreasing and differentiable and the random variable, $V = f(\mathbf{x})$ follows a density function $g(v)$. Denote

$$
\begin{aligned}
S(v) &= \{\mathbf{x} : \mathbf{x} \in \Re^n,\ h(\mathbf{x}'\mathbf{x}) \geq v\} \\
&= \{\mathbf{x} : \mathbf{x} \in \Re^n,\ \mathbf{x}'\mathbf{x} \leq h^{-1}(v)\},
\end{aligned}
$$

where $h^{-1}(\cdot)$ is the inverse function of $h(\cdot)$. Let $A(v)$ be the Lebesgue measure of $S(v)$. Then

$$
A(v) = \frac{2\pi^{n/2}}{n\Gamma(n/2)}[h^{-1}(v)]^{n/2} \tag{3.8}
$$

and

$$
A'(v) = \frac{\pi^{n/2}}{\Gamma(n/2)}[h^{-1}(v)]^{n/2-1}[h^{-1}(v)]'. \tag{3.9}
$$

The density of V is given by (Troutt, 1993) as

$$
g(v) = -\frac{\pi^{n/2}}{\Gamma(n/2)}v[h^{-1}(v)]^{n/2-1}[h^{-1}(v)]' \tag{3.10}
$$

Therefore the density of $W = V/f^*$ is given by

$$
p(w) = -\frac{f^*\pi^{n/2}}{\Gamma(n/2)}w[h^{-1}(f^*w)]^{n/2-1}[h^{-1}(f^*w)]', \tag{3.11}
$$

where $f^* = \max V = \max_{\mathbf{x}\in\Re^n} f(\mathbf{x}) = h(0)$. Using the general formulae (3.8)–(3.11), the multivariate vertical pdfs of several subclasses of spherically symmetric distributions can be shown explicitly as in the following sections.

3.4.2　*Multivariate Normal Distribution*

Let $\mathbf{x}$ be an n-dimensional random vector following a multivariate normal distribution with density function

$$f(\mathbf{x}) = (2\pi)^{-n/2} \exp\left(-\frac{1}{2}\mathbf{x}'\mathbf{x}\right), \quad \mathbf{x} \in \Re^n.$$

Consider the random variable $V = f(\mathbf{x})$. Writing $f^* = (2\pi)^{-n/2}$ and

$$h(t) \quad = \quad (2\pi)^{-n/2} \exp\left(-\frac{1}{2}t\right).$$

We have

$$h^{-1}(v) \quad = \quad -2\ln[(2\pi)^{n/2}v], \quad [h^{-1}(v)]' = -\frac{2}{v},$$

$$A(v) \quad = \quad \frac{2\pi^{n/2}}{n\Gamma(n/2)}[-2\ln((2\pi)^{n/2}v)]^{n/2},$$

$$g(v) \quad = \quad -\frac{2\pi^{n/2}}{\Gamma(n/2)}[-2\ln((2\pi)^{n/2}v)]^{n/2-1}, \quad 0 < v < f^*,$$

and

$$p(w) = \frac{2f^*}{\Gamma(n/2)}[-2\ln w]^{n/2-1}, \quad 0 < w < 1.$$

Let $Z = -2\ln W$, then Z has a density function

$$q(z) = \frac{1}{2^{n/2}\Gamma(n/2)} z^{n/2-1} \exp\left(-\frac{z}{2}\right), \quad 0 < z < \infty, \qquad (3.12)$$

i.e., $Z \sim \chi_n^2$. Thus,

$$Z = -2\ln W = -2\ln[(2\pi)^{n/2}V] = \mathbf{x}'\mathbf{x} \;\underline{d}\; R^2 \sim \chi_n^2,$$

where "$\overset{d}{=}$" means that the two sides have the same distribution. This relation is a basic property of the multivariate normal distribution $N_n(\mathbf{0}, \mathbf{I}_n)$. This result gives some ideas on the interrelation among the classical continuous distributions. Kotz, Fang and Liang (1997) consider the following two special cases:

Case 1. (c.f. Kotz and Troutt, 1996).
Let

$$A(v) = 2[-2\ln(\sqrt{2\pi}v)]^{1/2},$$

$$g(v) = 2[-2\ln(\sqrt{2\pi}v)]^{-1/2}, \quad 0 < v < \frac{1}{\sqrt{2\pi}},$$

and

$$p(w) = \sqrt{\frac{2}{\pi}}(-2\ln w)^{-1/2}, \quad 0 < w < 1.$$

Case 2. (c.f. Troutt, 1991).
Let

$$A(v) = -2\pi \ln(2\pi v),$$

$$g(v) = 2\pi, \quad 0 < v < \frac{1}{2\pi},$$

and

$$p(w) = 1, \quad 0 < w < 1,$$

where W is uniformly distributed on $(0,1)$.

Kotz, Fang and Liang (1997) further consider the Pearson Type VII distribution, where $f(\mathbf{x})$ is expressed as

$$f(\mathbf{x}) = C_n \left(1 + \frac{\mathbf{x}'\mathbf{x}}{m}\right)^{-N}, \quad \mathbf{x} \in \Re^n, \tag{3.13}$$

where $N > n/2$, $m > 0$, and $C_n = (m\pi)^{-n/2}\Gamma(N)/\Gamma(N - n/2)$. For this distribution,

$$h(t) = C_n \left(1 + \frac{t}{m}\right)^{-N},$$

$$h^{-1}(v) = m\left[\left(\frac{C_n}{v}\right)^{1/N} - 1\right]^{n/2}, \quad (h^{-1}(v))' = -\frac{m}{Nv}\left(\frac{C_n}{v}\right)^{1/N},$$

$$A(v) = \frac{2(m\pi)^{n/2}}{n\Gamma\left(\frac{n}{2}\right)}\left[\left(\frac{C_n}{v}\right)^{1/N} - 1\right]^{n/2},$$

$$g(v) \;=\; -vA'(v) = \frac{(m\pi)^{n/2}}{N\Gamma\left(\frac{n}{2}\right)} \left(\frac{C_n}{v}\right)^{1/N} \left[\left(\frac{C_n}{v}\right)^{1/N} - 1\right]^{n/2-1},$$

$$0 < v < f^* = C_n$$

$$W = V/f^* = V/C_n,$$

and

$$p(w) = \frac{1}{NB(\frac{n}{2}, N - \frac{n}{2})} w^{-1/N}(w^{-1/N} - 1)^{n/2-1}, \quad 0 < w < 1.$$

Setting $Z = W^{1/N}$, then Z has a density function

$$q(z) = \frac{1}{B(\frac{n}{2}, N - \frac{n}{2})} z^{N-n/2-1}(1 - z)^{n/2-1}, \quad 0 < z < 1, \tag{3.14}$$

That is, Z follows the Beta distribution with parameters $\frac{n}{2}$ and $N - \frac{n}{2}$. It is not difficult to obtain the density function of $\frac{R^2}{m}$:

$$\frac{1}{B(\frac{n}{2}, N - \frac{n}{2})} u^{n/2-1}(1 + u)^{-N}, \quad u > 0.$$

This is the density of the Beta II$(n/2, N - n/2)$ distribution. The multivariate t-distribution and the multivariate Cauchy distribution are two special cases of the above results:

First of all, let us consider the case of the multivariate t-distribution. Here the density function $f(\mathbf{x})$ is of the same form as that given by (3.13) with $N = n + m/2$ and $C_n = (m\pi)^{-n/2}\Gamma(n + m/2)/\Gamma(m/2)$. Thus,

$$A(v) = \frac{2(m\pi)^{n/2}}{(n + m)\Gamma\left(\frac{n}{2}\right)} \left[\left(\frac{C_n}{v}\right)^{2/(n+m)} - 1\right]^{n/2},$$

and

$$g(v) = -vA'(v) = \frac{2(m\pi)^{n/2}}{(n + m)\Gamma\left(\frac{n}{2}\right)} \left(\frac{C_n}{v}\right)^{2/(n+m)},$$

where

$$\left[\left(\frac{C_n}{v}\right)^{2/(n+m)} - 1\right]^{n/2-1}, \quad 0 < v < C_n.$$

In this case, $W = V/f^* = V/C_n$, $Z = W^{1/N} \sim \text{Beta}(n/2, m/2)$.

For the multivariate Cauchy distribution, $f(\mathbf{x})$ is also of the same form as that given by (3.13) with $m = 1, N = (n+1)/2$ and $C_n = \pi^{-n/2}\Gamma((n+1)/2)/\Gamma(1/2) = \Gamma((n+1)/2)/\pi^{(n+1)/2}$.

$$A(v) = \frac{2\pi^{n/2}}{n\Gamma(\frac{n}{2})}\left[\frac{1}{\pi}\left(\frac{\Gamma(\frac{n+1}{2})}{v}\right)^{2/(n+1)} - 1\right]^{n/2},$$

$$g(v) = -vA'(v) = \frac{2\pi^{n/2-1}}{(n+1)\Gamma(\frac{n}{2})}\left(\frac{\Gamma(\frac{n+1}{2})}{v}\right)^{2/(n+1)}.$$

$$\left[\frac{1}{\pi}\left(\frac{\Gamma(\frac{n+1}{2})}{v}\right)^{2/(n+1)} - 1\right], \quad 0 < v < \frac{\Gamma(\frac{n+1}{2})}{\pi^{\frac{n+1}{2}}}.$$

In this case, $W = V/f^* = V/C_n$ and $Z = W^{1/N} \sim \text{Beta}\,(n/2, 1/2)$.

3.4.3   *Tail Behaviour of the Multivariate Pearson Type II Distribution*

In Chapter 2, we defined a new random variable W, with pdf $p_W\,(w)\,,\ 0 < w < 1,\ w = v/m,\ v = f(x)$ and $m = \max\{f(x)\}$. The density $p_W\,(w)$ is called the normalized or modified VDR. We have also presented the way to indicate tail behaviour of a univariate pdf using its VDR and modified VDR. In this section, we present the tail behaviour of the multivariate Pearson Type II distribution using modified VDR.

The multivariate Pearson Type II distribution is defined as follows.

$$f(\mathbf{x}) = C_n(1 - \mathbf{x}'\mathbf{x})^m, \ C_n = \frac{\Gamma(\frac{n}{2} + m + 1)}{\pi^{n/2}\Gamma(m+1)}, \quad m > -1,$$

$$h(t) = C_n(1 - t)^m, \ 0 < t < 1,$$

$$h^{-1}(v) = 1 - \left(\frac{v}{C_n}\right)^{1/m},$$

$$A(v) = \frac{2\pi^{n/2}}{n\Gamma(\frac{n}{2})}\left[1 - \left(\frac{v}{C_n}\right)^{1/m}\right]^{n/2},$$

$$g(v) = -v'A(v) = \frac{\pi^{n/2}}{m\Gamma(\frac{n}{2})}\left(\frac{v}{C_n}\right)^{1/m}\left[1 - \left(\frac{v}{C_n}\right)^{1/m}\right]^{n/2-1}, \quad 0 < v < C_n,$$

$$W = V/f^* = V/C_n$$

and

$$p(w) = \frac{C_n\pi^{n/2}}{m\Gamma(\frac{n}{2})}w^{1/m}(1 - w^{1/m})^{n/2-1}, \quad 0 < w < 1.$$

Letting $Z = W^{1/m}$, we see that Z has a density function

$$q(z) = \frac{C_n\pi^{n/2}}{\Gamma(\frac{n}{2})}z^m(1 - z)^{n/2-1}$$

$$= \frac{1}{B(n/2, m+1)}z^m(1 - z)^{n/2-1}, \quad 0 < z < 1. \tag{3.15}$$

Thus, Z is distributed as the Beta distribution with parameters $\frac{n}{2}$ and $m+1$ and $R^2 \stackrel{d}{=} \mathbf{x}'\mathbf{x} \sim \text{Beta}(n/2, m+1)$, which is a well-known property of the Pearson Type II distribution.

3.4.4 Tail Behaviour of the Multivariate Spherically Symmetric Logistic Distribution

The density function of the multivariate logistic distribution is given by

$$f(\mathbf{x}) = \frac{C_n \exp(-\mathbf{x}'\mathbf{x})}{(1 + \exp(-\mathbf{x}'\mathbf{x}))^2}, \quad \mathbf{x} \in \Re^n,$$

where

$$C_n^{-1} = \int_{R^n} \frac{\exp(-\mathbf{x}'\mathbf{x})}{(1 + \exp(-\mathbf{x}'\mathbf{x}))^2}d\mathbf{x} = \frac{\pi^{n/2}}{\Gamma(\frac{n}{2})}\int_0^\infty t^{n/2-1}\frac{\exp(-t)}{(1 + \exp(-t))^2}dt$$

and

$$h(t) = \frac{C_n \exp(-t)}{(1 + \exp(-t))^2}, \quad t > 0.$$

Considering

$$V \;=\; h(\mathbf{x}'\mathbf{x}),$$

$$S(v) \;=\; \left\{ \mathbf{x} : \mathbf{x} \in \Re^n, \; \frac{C_n \exp(-\mathbf{x}'\mathbf{x})}{(1 + \exp(-\mathbf{x}'\mathbf{x}))^2} \geq v \right\}$$

$$= \left\{ \mathbf{x} : \mathbf{x} \in \Re^n, \; -\ln \frac{1 - \frac{2v}{C_n} + \sqrt{1 - \frac{4v}{C_n}}}{\frac{2v}{C_n}} \leq \mathbf{x}'\mathbf{x} \right.$$

$$\left. \leq -\ln \frac{1 - \frac{2v}{C_n} - \sqrt{1 - \frac{4v}{C_n}}}{\frac{2v}{C_n}} \right\}, \quad \text{for } 0 < v < \max V = \frac{C_n}{4}.$$

Let, as above, $A(v)$ be the Lebesgue measure of $S(v)$. Then

$$A(v) = \frac{2\pi^{n/2}}{n\Gamma(\frac{n}{2})} \left[\ln \frac{1 - \frac{2v}{C_n} + \sqrt{1 - \frac{4v}{C_n}}}{\frac{2v}{C_n}} \right]^{n/2}.$$

The density of V is,

$$g(v) \;=\; -vA'(v) = \frac{2\pi^{n/2}}{\Gamma(\frac{n}{2})} \left[\ln \frac{1 - \frac{2v}{C_n} + \sqrt{1 - \frac{4v}{C_n}}}{1 - \frac{2v}{C_n} - \sqrt{1 - \frac{4v}{C_n}}} \right]^{n/2-1} \cdot \sqrt{1 - \frac{4v}{C_n}}$$

$$= \frac{2\pi^{n/2}}{\Gamma(\frac{n}{2})} \left[2\ln \frac{1 - \frac{2v}{C_n} + \sqrt{1 - \frac{4v}{C_n}}}{\frac{2v}{C_n}} \right]^{n/2-1} \cdot \sqrt{1 - \frac{4v}{C_n}}, \qquad (3.16)$$

$$0 < v < f^* = \max V = \frac{C_n}{4}.$$

Then the random variable $W = V/f^* = 4V/C_n$ has a pdf,

$$p(w) \;=\; \frac{\pi^{n/2} C_n}{2\Gamma(\frac{n}{2})} \left[\ln \frac{1 - \frac{w}{2} + \sqrt{1 - w}}{1 - \frac{w}{2} - \sqrt{1 - w}} \right]^{n/2-1} \cdot \sqrt{1 - w}$$

$$= \frac{\pi^{n/2} C_n}{2\Gamma(\frac{n}{2})} \left[2\ln \frac{1 - \frac{w}{2} + \sqrt{1 - w}}{\frac{w}{2}} \right]^{n/2-1} \cdot \sqrt{1 - w}, \quad 0 < w < 1.$$

For $n = 2$, $C_2 = \pi^{-1}$, and $p(w) = 1/2 \cdot (1-w)^{1/2}$.
Also

$$R^2 = \ln \frac{1 - \frac{2v}{C_n} + \sqrt{1 - \frac{4v}{n}}}{\frac{2v}{C_n}}.$$

Then the pdf of $Z = R^2$ is given by

$$q(z) = \frac{(2\pi)^{n/2} C_n}{\Gamma(\frac{n}{2})} z^{n/2-1} \exp(-z) \frac{[1 - \exp(-z)]^2}{[1 + \exp(-z)]^4}, \quad z > 0. \qquad (3.17)$$

3.4.5 *Multivariate Uniform Distribution on the Unit Sphere*

Here the density function is

$$f(\mathbf{x}) = \begin{cases} \frac{\Gamma(\frac{n}{2})}{(2\pi)^{n/2}}, & \mathbf{x} \in \Re^n, \ \mathbf{x}'\mathbf{x} = 1, \\ 0, & \text{otherwise.} \end{cases}$$

In this case,

$$h(t) = \begin{cases} \frac{\Gamma(n/2)}{(2\pi)^{n/2}}, & |t| = 1, \\ 0, & \text{otherwise.} \end{cases} \qquad A(v) = \begin{cases} \frac{(2\pi)^{n/2}}{\Gamma(n/2)}, & 0 \le v \le \frac{\Gamma(n/2)}{(2\pi)^{n/2}}, \\ 0, & v > \frac{\Gamma(n/2)}{(2\pi)^{n/2}}. \end{cases}$$

Let $V = h(\mathbf{x}'\mathbf{x})$. Since $P\{\mathbf{x} : \mathbf{x}'\mathbf{x} = 1\} = 1$, we have the singular random variables V and W, which are constants almost surely with:

$$P\left\{ V = \frac{\Gamma(n/2)}{(2\pi)^{n/2}} \right\} = 1 \quad \text{and} \quad P\{W = 1\} = 1.$$

3.5 Tail Behavior of L_1 and L_p-Norm Symmetric Distributions

3.5.1 *L_1-Norm Symmetric Distributions*

Let us assume that the random vector, $\mathbf{x} = (x_1, \ldots, x_n)'$, has an L_1-norm symmetric distribution and possesses a density

$$f(\mathbf{x}) = h(\|\mathbf{x}\|_1), \qquad (3.18)$$

where $\mathbf{x} \in \Re_+^n$, $\|\mathbf{x}\|_1 = \sum_{i=1}^n x_i$ and $h(\cdot)$ is a strictly decreasing and differentiable function. Let

$$
\begin{aligned}
S(v) &= \{\mathbf{x} : \mathbf{x} \in \Re_+^n,\ h(\|\mathbf{x}\|_1) \geq v\} \\
&= \{\mathbf{x} : \mathbf{x} \in \Re_+^n,\ (\|\mathbf{x}\|_1) \leq h^{-1}(v)\},
\end{aligned}
$$

and

$$
A(v) = \text{ the Lebesgue measure of } S(v) = 1/n![h^{-1}(v)]^n. \tag{3.19}
$$

Thus, $A'(v) = \frac{1}{(n-1)!}[h^{-1}(v)]^{n-1}[h^{-1}(v)]'$. So $V = f(\mathbf{x}) = h(\|\mathbf{x}\|_1)$ (Troutt, 1991) has a density function

$$
g(v) = -vA'(v) = \frac{v}{(n-1)!}[h^{-1}(v)]^{n-1}[h^{-1}(v)]', \quad 0 < v < f^*, \tag{3.20}
$$

where $f^* = \max V = \max_{x \in \Re_+^n} f(\mathbf{x}) = h(0)$. It is also known that $\mathbf{x}$, which is assumed to have an L_1-norm symmetric distribution, has a stochastic representation $\mathbf{x} \overset{d}{=} R\,\mathbf{u}_n$, where the random variable $R > 0$ is independent of $\mathbf{u}_n$ and $\mathbf{u}_n$ is uniformly distributed on the simplex $\{\mathbf{x} : \mathbf{x} \in \Re_+^n, \|\mathbf{x}\|_1 = 1\}$. Thus $\mathbf{x}$ possesses a density if and only if R possesses a density (Fang, Kotz and Ng, 1990). In this case, the relation between the density, $b(t)$, of R and the density, $f(\mathbf{x}) = h(\|\mathbf{x}\|_1)$, of $\mathbf{x}$ is given by

$$
h(t) = \Gamma(n)t^{-n+1}b(t), \quad t > 0. \tag{3.21}
$$

Therefore, one can evaluate $A(v)$, $g(v)$ and $h(t)$ using (3.19) through (3.21). Kotz *et al.* (1997) have presented a collection of some of the subclasses of L_1-norm symmetric distributions with $h(\cdot)$ being strictly decreasing.

3.5.2   *L_p-Norm Symmetric Distributions*

A random vector, $\mathbf{x} = (x_1, \ldots, x_n)'$, is said to have an L_p-norm symmetric distribution if its density is given by

$$
h(\mathbf{x}) = C_n \prod_{i=1}^n x_i^{p-1} f(\|\mathbf{x}\|_p), \quad \mathbf{x} \in \Re_+^n,
$$

where $\|\mathbf{x}\|_p = \left(\sum_{i=1}^n x_i^p\right)^{1/p}$; and C_n is a constant, which depends only on n. Consider

$$
y_i = x_i^p, \ i = 1, \ldots, n. \tag{3.22}
$$

Then the random vector, $\mathbf{y} = (y_1, \ldots, y_n)'$, has an L_1-norm symmetric distribution with pdf given by

$$g(\mathbf{y}) = C_n f(\|\mathbf{y}\|_1), \quad \mathbf{y} \in \Re_+^n.$$

To generate the distribution of $\mathbf{x}$, we first generate the distribution of $\mathbf{y}$ by means of the result of Troutt (1993) and then the generation of $\mathbf{x}$ can be accomplished by (3.22).

3.6   Multivariate Burr, Pareto and Logistic Distributions

It is well known that multivariate Burr, Pareto, and Logistic distributions have the following density functions (See Chapter 9 of Johnson and Kotz, 1987) :

1) Burr distribution

$$
\begin{aligned}
f(\mathbf{x}) \;&=\; f(x_1, \ldots, x_n) \\
&=\; \frac{\Gamma(k+n)}{\Gamma(k)} \left[1 + \sum_{j=1}^{n} d_j x_j^{c_j} \right]^{-(k+n)} \prod_{j=1}^{n} d_j c_j x_j^{c_j - 1}, \quad \mathbf{x} \in \Re_+^n,
\end{aligned}
$$

    where $k > 0$, $d_j > 0$, $c_j > 0$ $(j = 1, \ldots, n)$ are constants.

2) Pareto distribution

$$
\begin{aligned}
f(\mathbf{x}) \;&=\; f(x_1, \ldots x_n) \\
&=\; \frac{\Gamma(a+n)}{\Gamma(a)} \left(\prod_{j=1}^{n} \theta_j \right)^{-1} \\
&\quad \left(\sum_{j=1}^{n} \theta_j^{-1} x_j - n + 1 \right)^{-(a+n)}, \quad x_j > \theta_j > 0,
\end{aligned}
$$

    where $a > 0$ and θ_j's are parameters.

3) Logistic distribution

$$
\begin{aligned}
f(\mathbf{x}) \;&=\; f(x_1, \ldots x_n) \\
&=\; \frac{\Gamma(\alpha+n)}{\Gamma(\alpha)} \exp\left(-\sum_{i=1}^{n} x_i \right) \left[\sum_{i=1}^{n} \exp(-x_i) + 1 \right]^{-(\alpha+n)},
\end{aligned}
$$

where $\mathbf{x} \in \Re^n$ and $\alpha > 0$ is a constant.

We need to carry out the following transformations in order to generate their vertical densities:

For the Burr distribution, let

$$y_j = d_j x_j^{c_j}, \quad j = 1, \ldots, n. \tag{3.23}$$

For the Pareto distribution, let

$$y_j = \theta_j^{-1} x_j, \quad j = 1, \ldots, n. \tag{3.24}$$

For the Logistic distribution, let

$$y_j = \exp(-x_j), \quad j = 1, \ldots, n. \tag{3.25}$$

Readers may refer to Table II in the article of Kotz *et al.* (1997) for full details of $g(v)$, $W = \frac{V}{c}$ and the distribution of Z.

3.7 VDR for the Multivariate Exponential Power Distribution

Kozubowski (2001) has generalized the result of Troutt (1991). The latter has showed that the ordinate density, $U = f(x_1, x_2)$, of a uncorrelated bivariate normal distribution is uniform, and posed a conjecture saying that the uniform density ordinate property holds for more general densities of the form

$$f(\mathbf{x}) = c_n \exp\{-(\mathbf{x}'\mathbf{x})^{n/2}\}, \quad \mathbf{x} \in \Re^n. \tag{3.26}$$

The proof of this conjecture can be found in the article by Kozubowski (2001), which we consider next.

Proposition 3.1   *Let $\mathbf{x}$ be a random vector in $\Re^n$ with pdf $f(\mathbf{x})$ given by (3.26) with $n \geq 1$. Then, the random variable, $U = f(\mathbf{x})$, has the uniform distribution on $(0, \ c_n)$.*

Proof.   Since the distribution of $\mathbf{x}$ is spherically symmetric with density, $f(\mathbf{x}) = h(\mathbf{x}'\mathbf{x})$, we have the well-known representation

$$\mathbf{x} \stackrel{d}{=} VT, \tag{3.27}$$

where T has the uniform distribution on the unit sphere $S_n = \{\mathbf{x} \in \Re^n : \mathbf{x}'\mathbf{x} = 1\}$, $V^2 = \mathbf{x}'\mathbf{x}$ is a positive random variable with pdf

$$u(y) = \frac{\pi^{n/2}}{\Gamma(n/2)} y^{n/2-1} h(y), \quad y > 0, \tag{3.28}$$

and T and V are independent (see, e.g., Muirhead, 1982, Theorem 1.5.5). Since in our case, $h(y) = c_n \exp\{-y^{n/2}\}$, the variable V^2 has a Weibull distribution with density function

$$u(y) = \frac{c_n \pi^{n/2}}{\Gamma(n/2)} y^{n/2-1} \exp\{-y^{n/2}\}, \quad y > 0, \tag{3.29}$$

(so that necessarily, $c_n = (n/2)\Gamma(n/2)/\pi^{n/2}$). Consequently, the random variable $(V^2)^{n/2} = (\mathbf{x}'\mathbf{x})^{n/2}$ has the standard exponential distribution, implying that $\exp\{-(\mathbf{x}'\mathbf{x})^{n/2}\}$ is standard uniform. This completes the proof.

Chapter 4

Applications of Multivariate VDR

Random Number Generation and The Vertical Strip Method

In this chapter, a method called the vertical strip (VS) method is proposed for generating non-uniform random variates with a given density. It can be considered as an improvement of the grid method (Devroye, 1986) as the VS method avoids setting up a directory to store information on big rectangles. Unlike the horizontal strip method that is based on the Riemann integral, the VS method is based on the Lebesgue integral and can be applied to unbounded densities or densities with infinite support. Applications of the VS method for generating random variates that follow the exponential distribution and normal density, are also given. Also, we discuss the generation of multivariate distributions by the VDR method in the second half of this chapter.

4.1　Generation of Non-Uniform Random Variates: An Overview

Ever since the feasibility of performing Monte Carlo experiments became a reality during World War II, vast literatures about generating samples from diverse distributions have grown insistently. The early works included Von Neumann (1951), Box and Muller (1958), Marsaglia (1964), and Ahrens and Dieter (1974). Generation of random numbers from a non-uniform distribution is usually achieved by means of a transformation to uniform variates.

There are a number of algorithms available for generating random numbers from certain common statistical distributions. These algorithms differ in many aspects; for example, they may differ in speed, level of accuracy and amount of computer storage space required, as well as coding structure. Probably, accuracy and speed are of the major concern to the end users. An algorithm may achieve high speed at the expense of accuracy.

The inverse cumulative distribution function (CDF) method is the most common method in random number generation. It involves inverting a function and applies directly only to univariate random variables. It is quite easy to use whenever an inverse CDF relationship exists between any two continuous random variables. Nevertheless, evaluating the inverse CDF even in closed form may be much slower than some alternative methods (Gentle, 1998). Using the inverse CDF for a general discrete distribution is essentially a table lookup operation (see Marsaglia, 1963; Norman and Canon, 1972 and Chen and Asau, 1974).

In recent times, the rejection-acceptance (RA) method has become one of the most useful techniques for sampling from the density, p, say, of a random variable X. The rejection-acceptance (RA) method makes use of realizations from another random variable Y, whose probability density $g(\cdot)$ is similar to the probability density of the target density, $p(\cdot)$. In fact, the random variable Y is so chosen that one can easily generate random deviates from it. The density, g, can be scaled to majorize p such that $cg(\cdot) \geq p(\cdot)$ using some constant c (See Devroye, 1986, 1987; Hörmann, 1994; and Hörmann and Deflinger, 1997). This method has been applied to generate non-uniform variates with continuous and discrete distributions. For generation of random variates from discrete distributions, one may refer to the article by Rajasekaran and Ross (1993) for details.

Wallace (1976) also suggested a method called " transformed rejection", which is a modification of the RA method. Marsaglia (1984) proposed a so-called " exact-approximation method". This method is very similar to the transformed rejection method. Devroye (1986) calls Marsaglia's method "almost exact inversion".

Recently, other methods to generate different types of distributions have been proposed. For example, Kinderman and Monahan (1977) suggested the ratio-of-uniforms method. Cheng and Feast (1979) used this method for gamma distributions, while Kinderman and Monahan (1980) applied the method to generate random deviates from the student t-distribution. Wakefield, Gelfand, and Smith (1991) generalized this method and applied it to some multivariate distributions. The ratio-of-uniforms method can also be applied to discrete distributions (see Ahrens and Dieter, 1991; Stadlober,

1990; Hörmann, 1994 and Hörmann and Deflinger, 1994).

For generation of more complicated distributions such as the density of the stationary distribution of a Markov chain, one may refer to the monographs by Gilks *et al.* (1996), Gamerman (1997) and Chen *et al.* (2000) for more details. Many of these algorithms such as the Metropolis algorithm (Metropolis *et al.*, 1953), Metropolis-Hastings algorithm (Hastings, 1970) and adaptive rejection sampling algorithm (Gilks and Wild, 1992) are hybrids of the rejection-acceptance (RA) method.

4.1.1 *The Vertical Strip Method*

One way to improve the RA method is to construct an appropriate envelope function of $p(x)$. However, a practical problem with this lies in finding a tight envelope function $M(x)$, that is $M(x) \geq p(x), \forall x \in \Re$. For the important class of log-concave univariate densities, Gilks and Wild (1992) proposed instead an adaptive rejection sampling method, but its computational cost was high. Kemp (1990) proposed the patchwork rejection algorithm to improve the RA method. Stadlober and Zechner (1999) applied this algorithm to unimodal distributions and suggested ways of rearranging the area below the density curve. A second RA sampling method is the grid method (Devroye, 1986), which is based on the fact that X has density p when (X, Y) is uniformly distributed under the density curve. The grid method is suitable for bounded densities with finite support. The area under the density curve is to be covered by a class A of small rectangles, which are either entirely contained within, or partly overlapped with the area under the density curve. By randomly choosing rectangles from A, the uniform random variate is first generated on the rectangle, and then the AR method is applied.

In what follows, we give a new construction of an envelope function for a given density, and propose a simple method called the vertical strip (VS) method, which is an improvement over the grid method, for generating random variates from the uniform distribution on a given sets. For comparison, the strip method employs a step-function as the envelope function based on the idea of the Riemann integral, while the vertical strip (VS) method uses the idea of the Lebesgue integral and vertical density representation (VDR).

4.2　Generation of the Uniform Distribution

Firstly, we give an explanation of the RA method in terms of Type II VDR (see Section 3.1 of Chapter 3).

Let the random variable X_1 have pdf $f(x)$ and let $D_2(f)$ be the set in $\Re^2$ defined by

$$D_2(f) = \{\mathbf{x}_2 = (x_1, x_2)' : 0 \leq x_2 \leq f(x_1)\}.$$

Let $\mathbf{X}_2 = (X_1, X_2)'$ be uniformly distributed on $D_2(f)$.

Then by lemma 3.1, we can apply the above results to generate the uniform distribution on triangular domains in $\Re^2$. Let us recall first the triangular distribution, which was studied in the 18th century, but has been the subject of only very few papers in the statistical literature over the last 50 years. It is noted that the Yawl distribution (Johnson and Kotz, 1999) recently proposed for financial applications, is a mixture of two triangular distributions.

Example 4.1　Triangular Distribution

In this example, we consider the uniform distribution on a triangular domain in $\Re^2$. Let $a \leq c \leq b$. Then the triangular density $\mathrm{TR}_{(a,b,c)}(\cdot)$, is defined as:

$$\mathrm{TR}_{(a,b,c)}(x) = \begin{cases} \frac{2}{b-a} \cdot \frac{x-a}{c-a}, & a \leq x \leq c; \\[2mm] \frac{2}{b-a} \cdot \frac{b-x}{b-c}, & c \leq x \leq b; \\[2mm] 0, & \text{otherwise} \end{cases}$$

Note that $D_2(\mathrm{TR}_{(a,b,c)})$ is a triangular domain of unit area. Devroye (1986) gave an algorithm for generating the uniform distribution on a triangular domain by considering the triangle as a convex polytope with three vertex points (Devroye, 1986, pp 568-569). In the sequel, we shall give another algorithm based on VDR.

Let (X, Y) be uniformly distributed on $D_2(\mathrm{TR}_{(a,b,c)})$. Then we have

(1) $X \sim \mathrm{TR}_{(a,b,c)}$;

(2) The density of Y is

$$
p_Y(y) = \begin{cases} \frac{b-a}{2} \cdot [2 - (b-a)y], & \text{for } 0 \le y \le \frac{2}{b-a}; \\[2mm] 0, & \text{otherwise.} \end{cases}
$$

(3) For given $Y = y$, the conditional distribution of X is uniform on the interval $D_{\text{TR}_{a,b,c}}(y) = (c - (c-a) \cdot \frac{2-(b-a)y}{2}, \ c + (b-c) \cdot \frac{2-(b-a)y}{2})$.

The following proposition provides an algorithm for generating the uniform distribution on the triangular domain, $D_2(\text{TR}_{(a,b,c)})$.

Proposition 4.1 Let U and V be independent random variables from $\mathcal{U}(0,1)$, then $(X, Y) = (c - (c-a)\sqrt{1-U} + (b-a)\sqrt{1-U}V, \ \frac{2(1-\sqrt{1-U})}{b-a})$ has the uniform distribution on $D_2(\text{TR}_{(a,b,c)})$ and $c - (c-a)\sqrt{1-U} + (b-a)\sqrt{1-U}V$ follows the triangular distribution, $\text{TR}_{(a,b,c)}(\cdot)$.

Proof. As in Figure 4.1,

$$
h = \frac{2}{b-a}, \quad l_y = \frac{b-a}{2} \cdot [2 - (b-a)y], \quad x = c - l_y \cdot \frac{c-a}{b-a}.
$$

Let (X, Y) be the random vector with uniform distribution on $D_2(\text{TR}_{(a,b,c)})$. Then the distribution function of Y is $F(y) = 1 - (1 - \frac{(b-a)y}{2})^2$. Thus, Y can be represented as $Y = \frac{2(1-\sqrt{1-U})}{b-a}$. Moreover, for given $Y = \frac{2(1-\sqrt{1-U})}{b-a} = y$, the conditional distribution of X is uniform on the interval,

$$
\left[c - l_y \cdot \frac{c-a}{b-a}, \ c + l_y \cdot \frac{b-c}{b-a} \right] = \left[c - (c-a)\sqrt{1-U}, \ c + (b-c)\sqrt{1-U} \right],
$$

so that X can be represented as

$$
X = c - (c-a)\sqrt{1-U} + (b-a)\sqrt{1-U}V.
$$

This completes the proof of the proposition.

In the case when $a = c$, then $a + (b-a)\sqrt{1-U}V$ has the triangular density $\text{TR}_{(a,b,a)}$. The best way to generate $\text{TR}_{(a,b,a)}$ is to use the patchwork rejection technique (Kemp, 1990). As illustrated in Figure 4.2, $D_2(\text{TR}_{(a,b,a)})$ is patched into the rectangle by rotation.

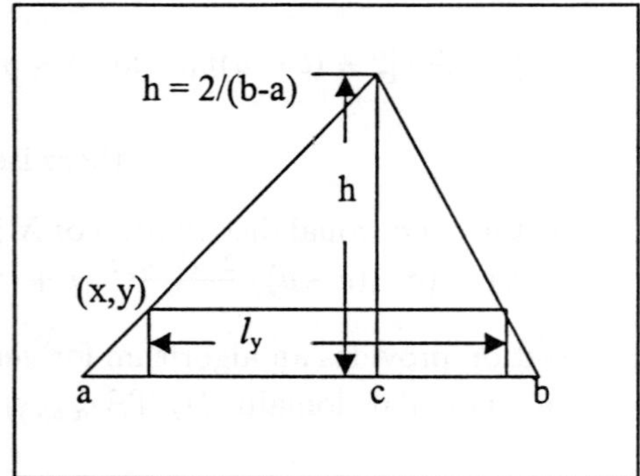

Figure 4.1: The triangular distribution

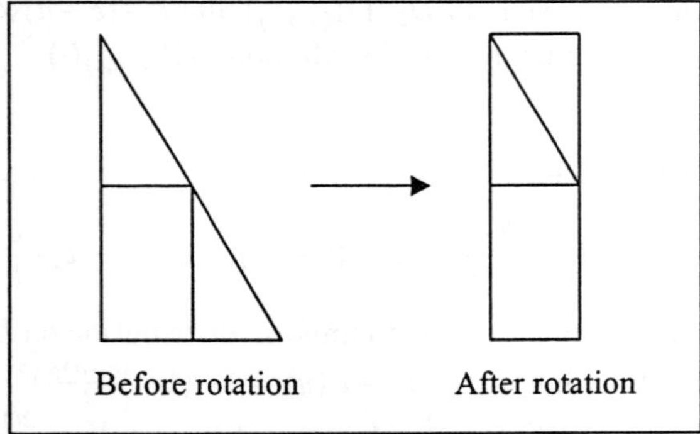

Figure 4.2: Patchwork into the rectangle

Thus, we have the following algorithm for generating random deviates that follow the triangular distribution.

Algorithm I

(1) Randomly generate U and V from $\mathcal{U}(0, 1)$.

(2) If $V \leq 0.5$ or $V \leq \frac{2-U}{2}$, set $Z = a + \frac{b-a}{2}U$. Else set $Z = \frac{a+b}{2} + \frac{(b-a)(1-U)}{2}$

(3) Deliver Z.

Example 4.2   The Trapezoidal Density

In this example, we propose a method for generating random variables having a trapezoidal density. Let $a \leq c \leq d \leq b$ and $h > 0$ be given. The trapezoidal density $\mathrm{TP}_{(a,b,c,d,h,[q])}(\cdot)$, is defined as

$$\mathrm{TP}_{(a,b,c,d,h,[q])}(z) = \begin{cases} \frac{h}{C}, & c \leq z \leq d, \\[2mm] \frac{q(z)}{C}, & a < z \leq c \ \text{ or } \ d \leq z \leq b, \\[2mm] 0, & \text{otherwise}, \end{cases} \qquad (4.1)$$

where $q(z)$ is a function on $[a, b]$ satisfying $q(z) \geq 0$, $q(c) = q(d) = h$, $q(a) = q(b) = 0$. Here, $C = h(d - c) + \int\limits_a^c q(z)dz + \int\limits_d^b q(z)dz$.

Suppose now that $q(\cdot)$ is a convex function satisfying:

$$q(z) \leq h \cdot \frac{z - a}{c - a}, \ a \leq z \leq c,$$

and

$$q(z) \leq h \cdot \frac{b - z}{b - d}, \ d \leq z \leq b.$$

Then it is clear that

$$\mathrm{TP}_{(a,b,c,d,h,[q])}(z) \ = (1 - \alpha - \beta)\frac{I_{(c,d)}(z)}{d-c} + \alpha \frac{q(z)}{\int\limits_a^c q(z)dz} I_{[a,c]}(z)$$

$$+ \beta \frac{q(z)}{\int\limits_d^b q(z)dz} I_{[d,b]}(z),$$

where

$$\beta = \frac{\int\limits_d^b q(z)dz}{C}, \qquad \alpha = \frac{\int\limits_a^c q(z)dz}{C}.$$

Denote the trapezoidal density by

$$\mathrm{TP}_{(a,b,c,d)}(z) = \frac{2(d-c)}{b-a+d-c}\frac{I_{[c,d]}(z)}{d-c} + \frac{c-a}{b-a+d-c}\mathrm{TR}_{(a,c,c)}(z)$$

$$+\frac{b-d}{b-a+d-c}\mathrm{TR}_{(d,b,d)}(z),$$

$$= \begin{cases} \frac{z-a}{c-a} \cdot \frac{2}{b-a+d-c}, & a \le z \le c, \\[2mm] \frac{2}{b-a+d-c}, & c \le z \le d, \\[2mm] \frac{b-z}{b-d} \cdot \frac{2}{b-a+d-c}, & d \le z \le b. \end{cases} \qquad (4.2)$$

Note that $\mathrm{TP}_{(a,b,c,c)}$ and $\mathrm{TP}_{(a,b,a,b)}$ are rectangular distributions. For $\gamma = \frac{h}{c} \cdot \frac{b-a+d-c}{2}$, $\gamma\mathrm{TP}_{(a,b,c,d)}$ is the envelope function of $\mathrm{TP}_{(a,b,c,d,h,[q])}$. Hence, we can generate $\mathrm{TP}_{(a,b,c,d,h,[q])}$ by the RA method and $D_2(\mathrm{TP}_{a,b,c,d,h,[q]})$ is patched into $\mathrm{TP}_{(s,t,s,t)}$, with $s = \frac{a+c}{2}, t = \frac{d+b}{2}$, by rotation. The algorithm to generate $\mathrm{TP}_{(a,b,c,h,[q])}$ is as follows:

Algorithm II
(1) Generate U from $\mathcal{U}(0,1)$, $U \leftarrow \frac{a+c}{2} + \frac{b-a+d-c}{2}U$;
(2) If $c \le U \le d$, then let $Z = U$, and deliver Z;
(3) Generate V from $\mathcal{U}(0,1)$. If $V \le \frac{U-a}{c-a}, U \le c$ and $hV \le q(U)$ or $V \le \frac{b-U}{b-d}, U > d$ and $hV \le q(U)$, then let $Z = U$, and deliver Z;
(4) If $U < c, V > \frac{U-a}{c-a}$ and $hV \le q(a+c-U)$, then let $Z = a+c-U$ and deliver Z;
(5) If $U > d, V > \frac{b-U}{b-d}$ and $hV \le q(b+d-U)$, then let $Z = b+d-U$, and deliver Z;
(6) Repeat steps 1-5 n times to generate n random variates from $\mathrm{TP}_{(a,b,c,h,[q])}$.

Algorithm II is based on the following lemma.

Lemma 4.1   *Let U be a random variable uniformly distributed on $[0,1]$ and let $\mathcal{I} = \{I_i = [a_i,b_i), i = 1,2,\cdots\}$ such that $[0,1] = \bigcup_{i=1}^{\infty} I_i$, $I_i \cap I_j = \emptyset$, $i \ne j$. Set $K = \sum_{i=1}^{\infty} iI_i(U)$ and $V = \frac{U-a_K}{b_K-a_K}$. Then V and K are independent of each other, and V is uniformly distributed on $[0,1]$.*

Proof.

$$P(V \leq v \mid K = k) \;=\; \frac{P(\{V \leq v\} \cap \{K = k\})}{P(K = k)}$$

$$=\; \frac{P(a_k \leq U \leq a_k + v(b_k - a_k))}{b_k - a_k}$$

$$=\; v$$

is independent of K and

$$P(V \leq v) \;=\; \sum_{i=1}^{\infty} P(V \leq v \mid K = k)P(K = k)$$

$$=\; v \sum_{k=1}^{\infty}(b_k - a_k)$$

$$=\; v.$$

The lemma is proven.

4.3 The Vertical Strip Method

Firstly, we give a brief recount of the RA method. Let the random variable Z have the pdf $p(z)$. Let $M(z)$ be a dominating function of p with

$$p(z) \leq M(z), z \in \Re \qquad \text{and} \qquad \int_{-\infty}^{\infty} M(z)dz = C < \infty.$$

We can write $p(z) = Cg(z)h(z)$, where $0 \leq g(z) = \frac{p(z)}{M(z)} \leq 1$, and $h(z) = \frac{M(z)}{C}$, as used in Von Neumann's representation. It is well known that the RA method can be used to generate the random variable X with density $p(x) = \frac{M(x)}{C}$ in the following ways:

(1) Generate a random variable U with uniform distribution on $[0, 1]$;
(2) If $U \leq \frac{p(X)}{M(X)}$ then return $Z = X$; else return to step (1).
(3) Repeat steps (1) and (2) until the desired number of random deviates are obtained.

It can be shown that Z follows density $p(\cdot)$. Indeed, using the notation in Section 4.2, we have $D_2(p) \subseteq D_2(M)$. If (Z, Y) follows the uniform distribution on $D_2(p)$, then Z follows the distribution of $p(\cdot)$ by Lemma 4.1.

One method of generating the random vector (Z, Y) is given below:

(1) Randomly generate (Z, Y) from the uniform distribution on $D_2(M)$;
(2) If $(X, V) \in D_2(p)$, then return $(Z, Y) = (X, V)$; else go to step (1).

Usually it is not difficult to find a function $M(\cdot)$ to generate the uniform distribution on $D_2(M)$. Let (X, V) be uniformly distributed on $D_2(M)$ with pdf $p_M(x, v) = \frac{1}{C} I_{D_2(M)}(x, v)$. It may be noted that:

(1) The pdf of X is

$$p_X(x) = \lim_{\Delta \to 0} \frac{P(x \le X \le x + \Delta)}{\Delta} = \lim_{\Delta \to 0} \frac{M(x)\Delta + o(\Delta)}{\Delta} = \frac{M(x)}{C};$$

(2) For given $X = x$, the conditional density $p_V(v \mid X = x)$ of V is

$$p_V(v \mid X = x) = \frac{p_M(x, v)}{p_X(x)} = \frac{1}{M(x)}.$$

Thus, the conditional density $p_V(v \mid X = x)$ of V is uniform on the interval $[0, M(x)]$, or $V = M(z)U$, where U is uniformly distributed on $[0, 1]$.

We note that $(X, V) \in D_2(p)$ is equivalent to $UM(X) \le p(X)$. Hence the RA method is essentially that of finding a domain $\mathcal{D} \supset D_2(p)$, and then randomly generating (X, V) from the uniform distribution on the domain $\mathcal{D}$ until $(X, V) \in D_2(p)$. The expected number of iterations is C and

$$P\{(X, V) \in D_2(p)\} = \frac{1}{C}.$$

4.3.1   *The Geometric Vertical Representation of a Density*

Let us begin with a definition of the geometric vertical representation of a density. Let $p(\cdot)$ be a continuous pdf on $\Re$ and suppose that there exists x_0 such that

(1) $p(x_0) = \sup\{p(x) : -\infty < x < \infty\}$;
(2) $p(\cdot)$ is increasing on $(-\infty, x_0]$ and is decreasing on $[x_0, \infty)$.

Let $\{h_i\}$ be a strictly monotone decreasing sequence of positive numbers such that

$$L_2(S_i) = 1 - \frac{1}{2^i}, i = 0, 1, 2, \cdots \tag{4.3}$$

where $S_i = \{(x, y) : (x, y) \in D_2(p), x \in D_p(h_i), y \geq h_i\}$. Hence, $\lim_{n \to \infty} h_n = 0$ and $\lim_{n \to \infty} L_2(S_n) = 1$. Then there exist two real sequences $\{a_i\}$ and $\{b_i\}$ satisfying

$$D_p(h_i) = [a_i, b_i], \ a_{i-1} \leq a_i < b_i \leq b_{i+1}, \ p(b_i) = p(a_i) = h_i, \ i = 0, 1, 2, \cdots \tag{4.4}$$

and

$$p(z) = \sum_{i=1}^{\infty} \frac{1}{2^i} \mathrm{TP}_{(a_i, b_i, a_{i-1}, b_{i-1}, h_{i-1} - h_i, [p_i])}(z), \tag{4.5}$$

where

$$p_i(z) = p(z) - h_i, a_i \leq z \leq a_{i-1}, \ \text{or} \ \ b_{i-1} \leq z \leq b_i.$$

By taking $\mathrm{TP}(a_i, b_i, a_i, b_i)$ as the envelope function of $\mathrm{TP}_{(a_i, b_i, a_{i-1}, b_{i-1}, h_{i-1} - h_i, [p_i])}$, we have the following vertical strip (VS) algorithm for generating variates with density, p.

Algorithm III

    (1) Randomly generate U from $\mathcal{U}(0, 1)$.
    (2) $U \leftarrow U + U$. If $U < 1$, then $U \leftarrow U + U$. If $U > 1$, then $U \leftarrow U - 1$;
    (3) $U \leftarrow a_k + U(b_k - a_k)$, if $a_{k-1} \leq U \leq b_{k-1}$, $Z = U$, return Z.
    (4) Randomly generate V from $\mathcal{U}(0, 1)$; if $h_k + (h_{k-1} - h_k)V \leq p(U)$, then $Z = U$, return Z
    (5) Repeat steps (1)-(4) until the desired number of random deviates are obtained.

If p has finite support, we can fix $h_i, i \geq 1$ by equal probability. If p has infinite support, we divide the interval, $[0, p_0]$, in accordance with a geometric distribution.

We denote $\mathrm{TP}_{(a_i, b_i, a_{i-1}, b_{i-1}, h_{i-1} - h_i, [p_i])}$ by TP_i. Let $\xi_i, i > 0$ be i.i.d. where ξ_i has density TP_i. If N follows the geometric distribution, $P(N = k) = \frac{1}{2^{k+1}}, k = 0, 1, 2, \cdots$, then ξ_N has density $p(\cdot)$. So, equation (4.5) may be considered as a *geometric vertical representation* of density,

$p(\cdot)$. This representation is valid for any pdf. Let $f(\cdot)$ be a pdf on $\Re^d$ and let $\{h_i\}$ be a positive sequence such that

$$L_{d+1}(S_i) = 1 - \frac{1}{2^i}, i = 0, 1, 2, \cdots$$

where $S_i = \{\mathbf{x}_{d+1} = (\mathbf{x}_d, x_{d+1}) : x_{d+1} \geq h_i, \mathbf{x}_{d+1} \in D_d(f)\}$. We have

$$f(\mathbf{x}_d) = \sum_{i=1}^{\infty} \frac{1}{2^i} f_i(\mathbf{x}_d),$$

where

$$f_i(\mathbf{x}_d) = \begin{cases} \frac{1}{2^i}(h_{i-1} - h_i), & \mathbf{x}_d \in D_f(h_{i-1}); \\[2mm] \frac{1}{2^i}(f(\mathbf{x}_d) - h_i), & \mathbf{x}_d \in D_f(h_i) \setminus D_f(h_{i-1}). \end{cases}$$

Hence, algorithm VS is also valid for higher dimensional densities.

4.3.2   *Generation of Random Variates from an Exponential Distribution*

Now we apply the VS method for generating random variates from two important distributions and make a comparison with classical methods. Ahrens and Dieter's (1988) method to generate random variates from the exponential distribution is fastest in Fortran and assembler languages. Performance measures such as the mean number of random numbers drawn from $\mathcal{U}(0,1)$ and the mean number of exponentiations are useful.

Let the random variable K follow the geometric distribution with parameter, $p = \frac{1}{2}$, and let V follow the truncated exponential distribution where

$$h(v) = 2e^{-v}, 0 \leq v \leq \ln 2.$$

Since

$$P(K = k) = \frac{1}{2^{k+1}}, \quad k = 0, 1, \cdots,$$

the random variable, $Z = K \ln 2 + V$, has density e^{-z}. Based on this result and the exact-approximation method, Ahrens and Dieter (1988) proposed the EA algorithm for generating the exponential distribution. Algorithm EA has the following desirable properties:

(1) The mean number of exponentiations is 2.983144 per 100 samples;

(2) The mean number of random numbers drawn from $\mathcal{U}(0,1)$ is $(0.98 + 0.02 * \frac{1}{1.511076})^{-1} = 1.06$ per sample.

With the VS method, we first divide the interval $[0, p(0)] = [0, 1]$ of density values according to the geometric series h_i, where $h_0 = 1$ and $h_i = \frac{1}{2^i}$ for $i > 0$. Setting $a_i = 0$, and $b_i = i \ln 2$, we have $e^{-b_i} = h_i$ and

$$e^{-x} = \sum_{i=0}^{\infty} I_{[i \ln 2, (i+1) \ln 2)} e^{-x} = \sum_{i=0}^{\infty} \frac{1}{2^{i+1}} e_i(x) \tag{4.6}$$

where

$$e_i(x) = \begin{cases} 2^{i+1} e^{-x}, & x \in [i \ln 2, (i+1) \ln 2); \\[2mm] 0, & \text{otherwise.} \end{cases}$$

It is easy to see that if the random variable, X, follows an exponential distribution, then $Y = X - i \ln 2 \sim h(v)$. In fact,

$$\begin{aligned} P(Y \leq y) \quad &= P(X \leq y + i \ln 2) \\ &= 2^{i+1}(e^{-i \ln 2} - e^{-i \ln 2 - y}) \\ &= 2 - 2e^{-y} = h(y), \ \forall y \in [0, \ln 2] \end{aligned}$$

Thus, it follows that $Z = K + V$ is exponentially distributed, as in Ahrens and Dieter's algorithm. Because e^{-x} is a convex function and $h(0.5 \ln 2) = \sqrt{2}$, we can take the envelope function of $h(\cdot)$ to be

$$\begin{aligned} \bar{h}(v) \quad &= \ln 2 \, \mathrm{TP}_1 + \left\{ (\sqrt{2} - 1) + \left[\tfrac{(\sqrt{2}-2)}{2} \right] \ln 2 \right\} \mathrm{TP}_2(v) \\[3mm] &\quad + \left\{ (2 - \sqrt{2}) - \tfrac{\sqrt{2}}{2} \ln 2 \right\} \mathrm{TP}_3(v) \\[3mm] &= \alpha_1 \mathrm{TP}_1(v) + \alpha_2 \mathrm{TP}_2(v) + \alpha_3 \mathrm{TP}_3(v) \end{aligned}$$

where

$$\mathrm{TP}_1(v) \quad = \mathrm{TP}_{(0,\ln 2,0,\ln 2)}(v)$$

$$\mathrm{TP}_2(v) \quad = \mathrm{TP}_{(0,\ln 2,0,0.5\ln 2)}(v)$$

$$\mathrm{TP}_3(v), \quad = \mathrm{TR}_{(0,0.5\ln 2,0)}(v)$$

are as shown in Figure 4.3 below.

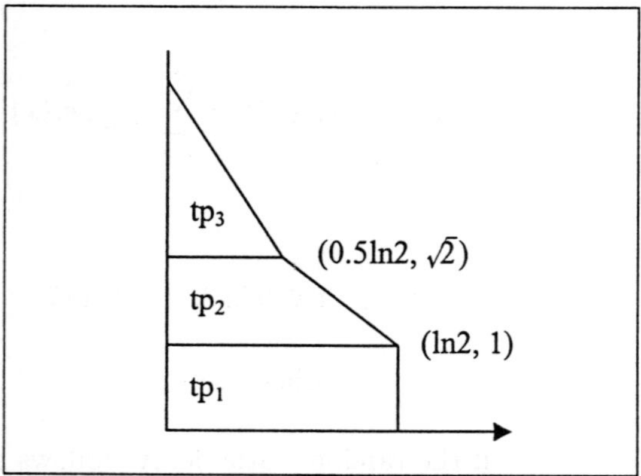

Figure 4.3: The Trapezoidal envelope function

The algorithm below is called the VS-E algorithm. It may be used to generate random variates from the exponential distribution.

Algorithm VS-E

Initialization constants:
(i) $w = \ln 2 = 0.6931471805$
(ii) $\alpha_1 = \ln 2$
(iii) $\alpha_2 = 0.2111954535$
(iv) $\alpha_3 = 0.0956573658$

Steps:

    (1) Randomly generate U from $\mathcal{U}(0,1)$. Set $k = 0$ initially.
    (2) While $U < 1$, set $k = k + 1$, and $U \leftarrow U + U$; else go to step 3.
    (3) $k \leftarrow k - 1$ and $U \leftarrow U - 1$
    (4) If $U < \alpha_1$, $U \leftarrow \frac{U}{\alpha_1}$; $Z = (U + k)w$, return Z.

(5) If $\alpha_1 \le U < \alpha_1 + \alpha_2$, $U \leftarrow \frac{U-\alpha_1}{\alpha_2}$; then,
(i) if $U \le \frac{2}{3}$, $Z = (0.75 \times U + k)w$, return Z;
(ii) if $U > \frac{2}{3}$; randomly generate V from $\mathcal{U}(0,1)$,
(a) if $V \le 2 - 1.5U$ and $(\sqrt{2} - 1)V \le 2e^{-U} - 1$, $Z = (U + k)w$, return Z;
else, let $U \leftarrow 1.5 - U$; and
(b) if $(\sqrt{2} - 1)(1 - V) \le 2e^{-U} - 1$, $Z = (U + k)w$, return Z.

(6) If $U > \alpha_1 + \alpha_2$, $U \leftarrow \frac{U-\alpha-1-\alpha_2}{\alpha_3}$; randomly generate V from $\mathcal{U}(0,1)$;
(i) if $V \le 1 - 0.5U$ and $(2 - \sqrt{2})V \le 2e^{-U} - \sqrt{2}$, $Z = (U + k)w$, return Z;
else if $V > 1 - 0.5U$, $U \leftarrow 1 - U$;
(ii) if $(2 - \sqrt{2})(1 - V) \le 2e^{-U} - \sqrt{2}$, $Z = (U + k)w$, return Z.
(7) Repeat until the desired number of random exponential deviates is obtained.

In brief, algorithm VS-E generates K from the geometric distribution with parameter $p = \frac{1}{2}$ and then generates a pair, (X, Y), from the uniform distribution on $D_2(\bar{h})$ until $(X, Y) \in D_2(h)$. Let $Z = X + K$. By Lemma 4.1, Z has density h. The acceptance probability is

$$
\begin{aligned}
P(\text{acceptance}) &= P((X,Y) \in D_2(h)) \\
&= L_2(D_2(h))/L_2(D_2(\bar{h})) \\
&= \frac{4}{(2\sqrt{2} + 3)\ln 2} \\
&= 0.9901,
\end{aligned}
$$

while

$$
\begin{aligned}
P\left(\begin{array}{c} \text{does not require exponentiation} \\ \text{computation} \end{array}\right) & \\
&= \frac{\int_0^{\frac{\ln 2}{2}} (1 - x + \frac{x^2}{2} - \frac{x^3}{6})dx + \int_{\frac{\ln 2}{2}}^{\ln 2} (1 - x + \frac{x^2}{2} - \frac{x^3}{6})dx}{L_2(C_p)} \\
&= \frac{0.2929538440541 + 0.205952809104}{0.5049947285771} \\
&= 0.98774625373996.
\end{aligned}
$$

Therefore, the mean number of exponentiations is $\frac{100(1-0.9877)}{0.98775} = 1.31$ per sample of 100 variates.

4.3.3   *Generation of Random Variates from a Normal Distribution*

There are many methods available to generate random variates from a standard normal distribution, such as the inverse transform method, the rectangle-wedge-tail method and the Box-Muller method. The RA or ratio-of-uniforms methods may be used to generate the half-normal distribution. The fastest one is Ahrens and Dieter's (1988) algorithm, which is essentially the Box-Muller method. In the sequel, we propose an algorithm using the VS method.

Let X be a random variable from the standard half normal distribution with pdf $p(x) = \sqrt{\frac{2}{\pi}}e^{-\frac{x^2}{2}}$. Let $p_i, i \geq 0$ and $b_i, i \geq 0$ be positive sequences such that

$$
\begin{aligned}
&L_2(S_i) = 1 - \tfrac{1}{2^i}, \\
&S_i = \{(x,y) : 0 \leq y \leq p(x), p(x) \geq p_i\}, \\
&p_i = p(b_i), i = 0, 1, \cdots
\end{aligned}
\tag{4.7}
$$

We can express the half normal density as

$$
p(x) = \sum_{i=1}^{\infty} \frac{1}{2^i} \mathrm{TP}_{(0,b_i,0,b_{i-1},[Tn_{(b_{i-1},b_i)}])}(x)
$$

where

$$
Tn_{(a,b)}(x) = p(x) - p(b), a \leq x \leq b.
$$

Owing to the fact that $p(x)$ is convex on $[1, \infty)$, the envelope function of $\mathrm{TP}_{(0,b_i,0,b_{i-1},p_{i-1}-p_i,[Tn_{(b_{i-1},b_i)}])}$ can be taken as $\mathrm{TP}_{(0,b_i,0,b_{i-1})}$. $\mathrm{TP}_{(0,b_1,0,b_1)}(\cdot)$ is one envelope function of $\mathrm{TP}_{(0,b_1,0,0,p(0)-p_1,[p_1])}$. We can therefore use the patchwork method to make it tighter. The maximum value of the density is $p(0) = \sqrt{\frac{2}{\pi}} = 0.7978845608$. The middle point of $p(0)$ and p_1 is $s = p(0) - \frac{p(0)-p_1}{2} = 0.5211739554$ and $p(r) = s$, $r = 0.9228573687$, $t = 2r - b_1 = 0.3077307798$.

As in Figure 4.4 below, domain 2 rotates the point around (r, s) to domain 3. Domain 3 does not overlap with domain 1 iff $\psi(y) = e^{\frac{-(y-r)^2}{2}} +$

$e^{\frac{-(y-r)^2}{2}} \leq 2e^{\frac{-r^2}{2}} = \psi(0)$. It is easy to see that $\psi'(0) = 0, \psi''(0) < 0$ and $\psi(0) = \sup\{\psi(y) : y \in \Re\}$. Hence $D_2(p_1)$ can be patched into $\mathrm{TP}_{(0,r,0,r)}$ by rotation.

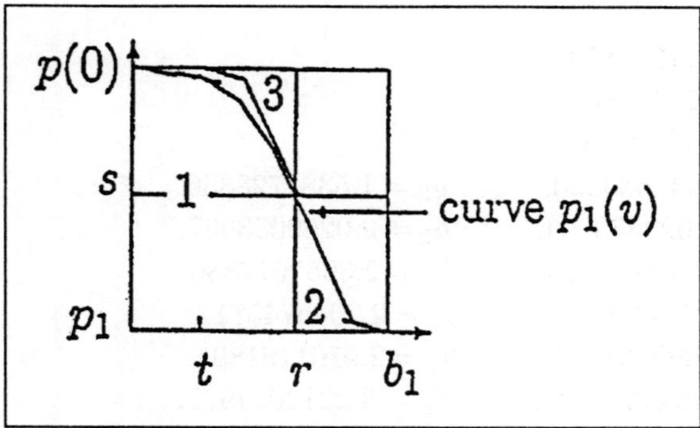

Figure 4.4: The half-normal curve and
the envelope function when $k = 1$.

In the following, we propose an algorithm to generate random variates from the standard half normal distribution using the VS method. We first compute p_i and b_i, $i = 1, 2, \ldots$ via the following relations. Let $g(y) = \sqrt{-2\ln(\sqrt{\frac{\pi}{2}}y)}$ be the inverse function of $p(\cdot)$ and

$$\int_{p_i}^{p_{i-1}} g(y)dy = \frac{1}{2^i}, \tag{4.8}$$

$$b_i = \sqrt{-2\ln(\sqrt{\frac{\pi}{2}}p_i)}, \tag{4.9}$$

where $p_0 = p(0) = \sqrt{\frac{\pi}{2}}$ and $b_0 = 0$.

The algorithm below is called the VS-N algorithm and may be used to generate random variates from the standard normal distribution.

Algorithm VS-N:

Initialization constants:

(i) $p_0 = p(0) = 0.7978845608.$

(ii) $b_0 = 0.0$

(iii) $s = 0.5211739554$

(iv) $r = 0.9228573687$

(v) $t = 0.3077307798.$

(vi) $p_1 = 0.2444633520,$ $b_1 = 1.5381728836.$

(vii) $p_2 = 0.1022879704,$ $b_2 = 2.0269053907.$

(viii) $p_3 = 0.0452524685,$ $b_3 = 2.3957072886.$

(iv) $p_4 = 0.0205802455,$ $b_4 = 2.7046745712.$

(x) $p_5 = 0.0095193841,$ $b_5 = 2.9761161891.$

(xi) $p_6 = 0.0044543803,$ $b_6 = 3.2212034975.$

(xii) $p_7 = 0.0021020169,$ $b_7 = 3.4464667745.$

(xiii) $p_8 = 0.0009983776,$ $b_8 = 3.6561147804.$

Steps:

(1) Randomly generate U from $\mathcal{U}(0,1)$. Set $k = 0$ initially.

(2) While $U < 1$, set $k = k + 1$, and $U \leftarrow U + U$; else if $U < 1$, go to step (3).

(3) $U \leftarrow U - 1$ and set $S = \mathrm{sgn}(U - 0.5)$. If $S = +1$, then $U \leftarrow U + U - 1$; else set $U \leftarrow U + U$.

(4) If $k = 1$, then randomly generate V from $\mathcal{U}(0,1)$.

    (a) If $rU \leq t$ and $p_1 + (p(0) - p_1)V \leq p(U)$, then $Z = rU$, return SZ.

    (b) If $rU > t$, $V \leq \frac{b_1 - U}{b_1 - t}$ and $p_1 + (p(0) - p_1)V \leq p(U)$, then $Z = rU$, return SZ.

    (c) If $p_1 + (p(0) - p_1)(2s - V) \leq p(U)$, then $U \leftarrow 2r - rU$, and $Z = rU$, return SZ.

(5) If $k > 1$ and $k \leq 8$, let $X = U(\frac{b_k + b_{k-1}}{2})$ and randomly generate V from $\mathcal{U}(0,1)$.

    (a) If $X \leq b_{k-1}$, then Z=X, return SZ.

    (b) If $V < \frac{b_k - U}{b_k - b_{k-1}}$, then

        i. if $p_k + (1 - V)(p_{k-1} - p_k) < p(U)$, then $Z = X$, return SZ;

ii. if $p_k + (1 - V)(p_{k-1} - p_k) \leq p(b_k + b_{k-1} - X)$, then $Z = b_k + b_{k-1} - X$, return SZ.

(6) If $k \geq 9$, we need to calculate p_k and b_k using equations (4.8) and (4.9) and go to step (5). However, the chance of having to evaluate p_k and b_k when $k \geq 9$ is negligible.

(7) Repeat until the desired number of random normal deviates is obtained.

For given $k \geq 2$, the probability of acceptance is

$$P(AC \mid k) = \frac{1}{2^{k-1}(b_{k-1} + b_k)(h_{k-1} - h_k)}$$

and the first four terms are listed below:

$$P(AC \mid k = 1) = \frac{0.5}{r(p(0) - p_1)} = 0.9791;$$
$$P(AC \mid k = 2) = 0.9864;$$
$$P(AC \mid k = 3) = 0.9911;$$
$$P(AC \mid k = 4) = 0.9934.$$

This method can also be applied to the gamma, beta and other continuous distributions. One possible factor that would affect the speed of computation might be the evaluations of p_i and b_i, but the chance of their required calculation for $k > 8$ is very small.

4.4 Computational Performance

In this section, we evaluate and compare the performance of the new VS-E and VS-N algorithms proposed in Section 4.3 with other existing algorithms and give a summary of our findings. For the generation of random variates from a standard exponential distribution, we compare the speed of the VS-E algorithm with those of the EA algorithm (Ahrens and Dieter, 1988) and SA algorithm (Ahrens and Dieter, 1972). We also compare the VS-N algorithm with the RA method by Kinderman and Ramage (1976) and the NA algorithm by Ahrens and Dieter (1988). These algorithms are all of similar nature, that is, generation of random deviates using either a ratio-of-uniforms method or a mixture with the RA method.

The implementations of all the programs were written in Fortran and run on a Unix Sun Sparc workstation. The average total CPU time used by each program for generating 10,000, 100,000 and 1 million random deviates for 100 repetitions are reported. In this way, we can compare the relative computational efficiencies of the algorithms. All $(0,1)$ uniform deviates used in each program were obtained by calling the IMSL (1991) Fortran subroutine RNUNF for standard precision and DRNUNF for double precision.

We present our computational results of the performance of each algorithm in Tables 4.1 to 4.4. Tables 4.1 and 4.2 are the results of standard exponential deviates for standard and double precision, respectively, while Tables 4.3 and 4.4 are the results of standard normal deviates for standard and double precision, respectively.

Table 4.1. Standard Exponential Distribution: Computational Performances of SA, EA and VS-E Algorithms (Standard Precision)

Methods	Average CPU Time		
No. of deviates generated	n=10,000	n=100,000	n= 1 million
SA method	0.10 sec.	0.53 sec.	4.69 sec.
EA method	0.09 sec.	0.42 sec.	4.32 sec
VS-E method	0.05 sec	0.36 sec	3.67 sec

Table 4.2. Standard Exponential Distribution: Computational Performances of SA, EA and VS-E Algorithms (Double Precision)

Methods	Average CPU Time		
No. of deviates generated	n=10,000	n=100,000	n= 1 million
SA method	0.14 sec.	0.57 sec.	4.91 sec.
EA method	0.11 sec.	0.48 sec.	4.85 sec.
VS-E method	0.06 sec	0.38 sec	3.85 sec

Table 4.3. Standard Normal Distribution: Computational Performances
of AR, NA and VS-N Algorithms (Standard Precision)

Methods	Average CPU Time		
No. of deviates generated	n=10,000	n=100,000	n= 1 million
RA method	0.15 sec.	0.79 sec.	7.82 sec.
NA method	0.17 sec.	0.93 sec.	9.35 sec.
VS-N method	0.05 sec.	0.53 sec.	5.06 sec.

Table 4.4. Standard Normal Distribution: Computational Performances
of RA, NA and VS-N Algorithms (Double Precision)

Methods	Average CPU Time		
No. of deviates generated	n=10,000	n=100,000	n= 1 million
RA method	0.17 sec.	0.85 sec.	8.29 sec.
NA method	0.20 sec.	0.97 sec.	10.32 sec.
VS-N method	0.06 sec.	0.58 sec.	5.38 sec.

As we can see from Tables 4.1-4.4, the computation speeds of the VS methods are relatively faster.

Thus, computational experience indicates that our newly proposed VS methods based on the concept of the Lebesgue integral and VDR are sound and viable. They can also be applied to generate gamma, beta and other common continuous distributions. Moreover, these are very easy to implement and considerably faster than some other existing methods.

4.5 Generation of Multivariate Distributions

In the previous sections, we have introduced two algorithms for generating the standard exponential and normal distributions respectively by the VS method. This is a major application of VDR in Monte Carlo simulation. In theory, we can extend this method to generate other common univariate distributions and it has been proved empirically that this method is computationally more efficient than the other existing algorithms. For generation of multivariate distributions, there exist several approaches, for example, the transformation method, RA method and its several improvements, such as ratio-of-uniforms method and the exact approximation method, stochastic representation and the conditional distribution method. In this section, we shall introduce the method of VDR for the generation of multivariate distributions.

Fang *et al.* (2001) have introduced the so-called Type II VDR. Based on the Type II VDR method, we can generate the uniform distribution on a manifold using the idea of the multivariate vertical density representation (MVDR).

Now let the random vector $\mathbf{X} = (X_1, \ldots, X_n)$ follows the uniform distribution on a Lebesgue measurable region D_n in $\Re^n$, satisfying $0 < L_n(D_n) < \infty$, where $L_n(D_n)$ is the Lebesgue measure of D_n. Set

$$D_{n-1}(v) = \{\mathbf{x}_{n-1} = (x_1, \ldots, x_{n-1}) : (\mathbf{x}_{n-1}, v) \in D_n\}. \tag{4.10}$$

Then

$$L_n(D_n) = \int_{-\infty}^{\infty} L_{n-1}(D_{n-1}(v))dv, \tag{4.11}$$

and the pdf of $\mathbf{X}$ is given by

$$p_{\mathbf{x}}(\mathbf{x}) = \begin{cases} (1/L_n(D_n)), & \text{if } \mathbf{x} \in D_n, \\ 0, & \text{if } \mathbf{x} \notin D_n. \end{cases} \tag{4.12}$$

According to the above definitions, we have the following theorem (Theorem 1 of Fang *et al.*, 2001).

Theorem 4.1 *Let D_n be a Lebesgue measurable set in $\Re^n$ such that $0 < L_n(D_n) < \infty$ and $L_{n-1}(D_{n-1}(v))$ is continuous in v. Then, $\mathbf{X} = (X_1, \ldots, X_n)$ has the uniform distribution on D_n iff*

(1) *The marginal pdf of X_n is $(L_{n-1}(D_{n-1}(v))/L_n(D_n))$;*
(2) *For given $X_n = v$, the conditional distribution of $\mathbf{X}_{n-1} = (X_1, \ldots, X_{n-1})$ is the uniform distribution on $D_{n-1}(v)$.*

Proof. By the definition of density

$$p_{x_n}(x) = \lim_{\Delta \to 0} \frac{P(x \leq X_n \leq x + \Delta)}{\Delta}.$$

Since $L_{n-1}(D_{n-1}(\cdot))$ is continuous, we have

$$
\begin{aligned}
P(x \leq X_n \leq x + \Delta) &= \frac{L_n(\{x \leq X_n \leq x + \Delta\} \cap D_n)}{L_n(D_n)} \\
&= \frac{1}{L_n(D_n)} \int_x^{x+\Delta} L_{n-1}(D_{n-1}(s))ds \\
&= \frac{L_{n-1}(D_{n-1}(x^*))}{L_n(D_n)} \Delta
\end{aligned}
$$

where x^* is a point in the closed interval $[x, x + \Delta]$. Hence, the pdf of X_n is given by

$$
p_{x_n}(x) = \lim_{\Delta \to 0} \frac{P(x \leq X_n \leq x + \Delta)}{\Delta} = \frac{L_{n-1}(D_{n-1}(x))}{L_n(D_n)}.
$$

For given $X_n = v$, the conditional pdf of $\mathbf{X}_{n-1} = (X_1, \ldots, X_{n-1})$ is

$$
\begin{aligned}
p_{(\mathbf{x}_{n-1}|x_{n=v})}(x_{n-1}) &= \frac{1}{L_n(D_n)} \cdot \frac{L_n(D_n)}{L_{n-1}(D_{n-1}(v))} I_{D_n}(\mathbf{x}_{n-1}, v) \\
&= \frac{I_{D_{n-1}(v)}(x_{n-1})}{L_{n-1}(D_{n-1}(v))},
\end{aligned}
$$

where I_A is the indicator function of set A. This completes the proof.

Theorem 4.1 can be generalized. For a given integer $0 < k < n$, put

$$
D_k(\mathbf{v}_{|k|}) = \{\mathbf{x}_k = (x_1, \ldots, x_k) : (\mathbf{x}_k, \mathbf{v}_{|k|}) \in D_n\},
$$

where $\mathbf{v}_{|k|} = (v_{k+1}, \ldots, v_n)$.

Corollary 4.1 *Let a random vector* $\mathbf{X}_n = (X_1, \ldots, X_n)$ *satisfy the following two conditions:*

(1) *The pdf of* X_n *is* $(L_{n-1}(D_{n-1}(x))/L_n(D_n))$.
(2) *For given* $X_{[k]} \equiv (X_{k+1}, \ldots, X_n) = v_{[k]}$, *the conditional pdf of* X_k *is*

$$
(L_{k-1}(D_{k-1}(x_k, v_{[k]}))/L_k(D_k(v_{[k]}))), k = n - 1, \ldots, 1.
$$

Then the distribution of $\mathbf{X}_n$ *is the uniform distribution* $U(D_n)$ *and*

$$
L_{k+1}(D(v_{[k+1]})) = \int_{-\infty}^{\infty} L_k(D_k(v_{[k]}))dv_{k+1}, \quad k = 1, \ldots, n - 1.
$$

The following examples are given in the article by Fang *et al.*, (2001).

Example 4.3 Example 1 Fang *et al.* (2001): Uniform distribution on a ball.

Let $B_n(r) = \{\mathbf{x} \in \Re^n : \mathbf{x}'\mathbf{x} \le r^2\}$ be a ball in $\Re^n$ and $B_n = B_n(1)$ be the unit ball. Tashiro (1977) and Fang and Wang (1994) described stochastic representation for $\mathbf{X} \sim U(B_n)$. We now apply Theorem 4.1 to $D_n = B_n$ and provide another method for generating the uniform distribution $U(B_n)$. Note that $D_{n-1}(v) = B_{n-1}(\sqrt{1-v^2})$ and

$$
\begin{aligned}
L_{n-1}(D_{n-1}(v)) &= L_{n-1}(B_{n-1}(\sqrt{1-v^2})) \\
&= \frac{2\pi^{((n-1)/2)}}{(n-1)\Gamma((n-1)/2)}\left(\sqrt{1-v^2}\right)^{n-1}.
\end{aligned}
$$

If $\mathbf{X}_n = (X_1,\ldots,X_n) \sim U(B_n)$, then from Theorem 4.1, the marginal distribution function of X_n is given by

$$
\begin{aligned}
F_n(x_n) &= \int_{-1}^{x_n} \frac{L_{n-1}(D_{n-1}(v))}{L_n(D_n)}\,dv \\
&= \left[\int_{-1}^{0} + \int_{0}^{x_n}\right] \frac{n\Gamma(n/2)}{(n-1)\Gamma((n-1)/2)\sqrt{\pi}}(\sqrt{1-v^2})^{n-1}dv \\
&= \frac{1}{2} + \frac{1}{2}sign(x_n)\frac{\Gamma((n+2)/2)}{\Gamma((n+1)/2)\sqrt{\pi}}\int_{0}^{x_n^2}(1-s)^{((n-1)/2)}s^{-1/2}ds \\
&= \frac{1}{2} + \frac{1}{2}sign(x_n)beta\left(x_n^2; \frac{n+1}{2}, \frac{1}{2}\right),
\end{aligned}
$$

where $beta(\cdot; \alpha, \beta)$ is the beta distribution function with parameters, α, β and

$$
sign(x) = \begin{cases} 1, & \text{if } x > 0; \\ 0, & \text{if } x = 0; \\ -1, & \text{if } x < 0. \end{cases}
$$

From Theorem 4.1, we have the following algorithm to generate the uniform distribution on B_n. Let $U_1,\ldots,U_n$ be random numbers and $beta^{-1}(\cdot)$ be the inverse function of the beta cumulative distribution function $beta(\cdot; \alpha, \beta)$. Set

$$
X_k = sign\left(U_k - \frac{1}{2}\right)\sqrt{beta^{-1}\left(U_k; \frac{k+1}{2}, \frac{1}{2}\right)}
$$

$$\times \left(1 - \sum_{i=k+1}^{n} X_i^2\right), \qquad k = 1, \ldots, n,$$

where the summation, $\sum_m^n(\cdot)$, is 0 if $m > n$, then $X_1, \ldots, X_n$ is uniformly distributed on the ball B_n.

Example 4.4 (Example 2 of Fang, Yang and Kotz 2001). The Uniform Distribution on a Convex Set

Consider the convex set $V_n = \{\mathbf{X} = (X_1, \ldots, X_n) : \sum_{i=1}^{n} |X_n| \leq 1\}$ (See Figure 4.5). It is obvious that

$$\begin{aligned} V_n(v) &= \{\mathbf{X} = (X_1, \ldots, X_n) : X_n + v, \ \mathbf{X} \in V_n\} \\ &= \{\mathbf{X} = (\mathbf{X}_{n-1}, v) : \mathbf{X}_{n-1} = (X_1, \ldots, X_{n-1}) \in V_{n-1, 1-|v|}\}, \end{aligned}$$

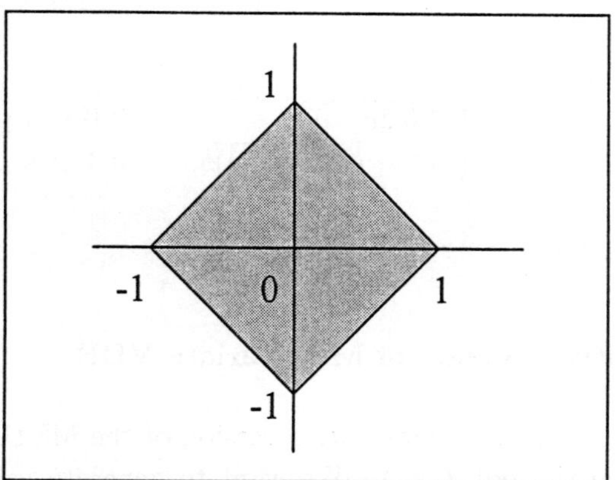

Figure 4.5: The uniform distribution on a convex set

where $V_{m,a} = \{\mathbf{X} = (X_1, \ldots, X_m) : \sum_{i=1}^{m} |X_i| \leq a\}$. It is known (Fleming, 1997) that the volume of $V_{m,a}$ is

$$L_m(V_{m,a}) = \frac{2^m}{m!} a^m.$$

Suppose the random vector $\mathbf{X}_n \sim U(V_n)$. From Theorem 4.1, the pdf of $\mathbf{X}_n$ is given by

$$p_{\mathbf{X}_n}(v) = \frac{2^{n-1}/(n-1)!}{2^n/n!}(1 - |v|)^{n-1} = \frac{n}{2}(1 - |v|)^{n-1}, \quad -1 \leq v \leq 1$$

and its distribution function is

$$
F_{X_n}(v) = \begin{cases} 1, & \text{if } v \geq 1, \\ 1 - \frac{1}{2}(1-v)^n, & \text{if } 0 \leq v \leq 1 \\ \frac{1}{2}(1+v)^n, & \text{if } -1 \leq v \leq 0 \\ 0, & \text{if } v \leq -1. \end{cases}
$$

Therefore, we have an algorithm for generating the uniform distribution on V_n as follows:

(1) Generate random numbers $U_1, \ldots, U_n$.
(2) Calculate

$$
X_{n-k} = G\left(n-k, 1 - \sum_{i=n-k+1}^{n} |X_i|, U_{n-k}\right), \qquad k = 0, 1, \ldots, n-1,
$$

where

$$
G(k, b, u) + \begin{cases} (\sqrt[k]{2v} - 1)b, & \text{if } 0 \leq u \leq 1/2, \\ (1 - \sqrt[k]{2(1-v)})b, & \text{if } 1/2 \leq u \leq 1. \end{cases}
$$

Deliver $X_1, \ldots, X_n$.

4.6   Alternative Version of Multivariate VDR

In this section, we give an alternative version of the MVDR. Let random vector $\mathbf{X}_n$ have the pdf $f(\mathbf{x}_n)$. We want to generate variates from this distribution. From the theory of the AR method and the idea of MVDR, we have the following theorem (Theorem 2 of Fang *et al.*, 2001):

Theorem 4.2   *Let $\mathbf{X}_{n+1} = (\mathbf{X}_n, X_{n+1})$ have the uniform distribution on $D_{n+1}(f)$, where $D_{n+1}(f) = \{\mathbf{x} = (x_1, \ldots, x_{n+1} = (\mathbf{x}_n, x_{n+1} : 0 \leq x_{n+1} \leq f(\mathbf{x}_n)\}$. Then we have*

(1) The pdf of $\mathbf{X}_n$ is $f(x)$.
(2) The pdf of X_{n+1} is $p(x) = L_n(S_f(x))$.
(3) For given $X_{n+1} = v$, $0 \leq v \leq \sup_x f(\mathbf{x})$, the conditional distribution of $\mathbf{X}_n$ is the uniform distribution, $U(S_f(v))$.

In fact, Theorem 4.2 can imply the VDR proposed by Troutt (1991). Note that using the notations of Theorem 4.2, we have

$$\begin{aligned}
\{(\mathbf{x}_n, x_{n+1}) &: & f(\mathbf{x}_n) \leq v, (\mathbf{x}_n, x_{n+1}) \in D_{n+1}(f)\} \\
&= & \{(\mathbf{x}_n, x_{n+1}) \in D_{n+1}(f); \mathbf{X}_{n+1} \leq v\}\backslash \\
\{(\mathbf{x}_n, x_{n+1}) &\in & D_{n+1}(f) : x_{n+1} \leq v\} \\
\cap \{(\mathbf{x}_n, x_{n+1}) &\in & D_{n+1}(f) : f(\mathbf{x}_n) > v\}.
\end{aligned}$$

It follows that

$$P(f(\mathbf{X}_n) \leq v) = \int_0^v L_n(S_f(x)dx - vL_n(S_f(v)).$$

Also, if $L_n(S_f(\mathbf{x}))$ is differentiable, then the density of $V = f(\mathbf{X}_n)$ is

$$p_V(v) = \frac{d}{dv}P(f(\mathbf{X}_n) \leq v) = -v\frac{d}{dv}L_n(S_f(v)),$$

which results in Theorem 1 of Troutt (1991). We now give some applications of the above theory. First, we provide an intuitive interpretation of Theorem 4.2.

Example 4.5 (Example 3 of Fang *et al.*, 2001). Generating A Univariate Distribution

Let the random variable X have a density $f(x)$. If we can generate (X_1, X_2) that follows the uniform distribution on $D_2(f) = \{(x_1, x_2) : x_2 \leq f(x_1)\}$ then X_1 and X_2 have the pdfs, $f(x)$, and $L_1(S_f(v))$, respectively. The distribution function of X_2 is

$$F_2(v) = \int_0^v L_1(S_f(s))ds, \qquad 0 \leq v \leq \sup_x f(x),$$

from Theorem 4.2. The conditional distribution of X_1 given $X_2 = v$ is the uniform distribution on $S_f(v)$.

Suppose that $f(x)$ is continuous, increasing on $(-\infty, a]$ and decreasing on $[a, \infty)$ for some a. It is obvious that $M = \max_x f(x) = f(a)$ and $F_2(v)$ is the area of the region $E = \{(x_1, x_2) : x_2 \leq v\} \cap D_2(f)$. From Figure 4.6, it is easy to see that

$$F_2(v) = 1 - F(a_u) + F(a_1) + f(a_u)(a_u - a_1), \tag{4.13}$$

and

$$S_f(v) = [a_l, a_u], \tag{4.14}$$

where

$$v = f(a_l) = f(a_u), \qquad a_l < a_u, \tag{4.15}$$

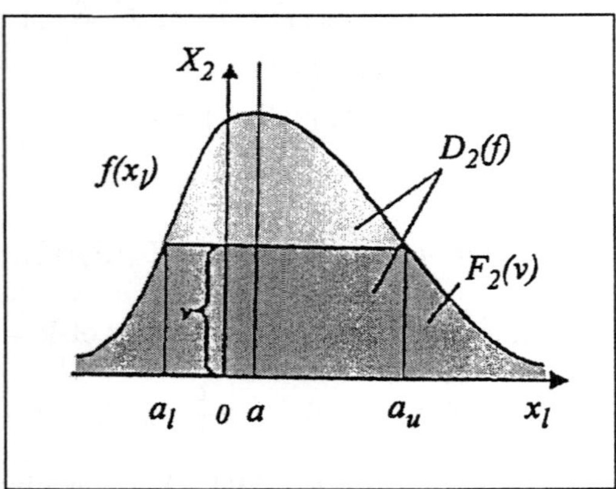

Figure 4.6: The set $D_2(f)$ and $F_2(v)$.

and $F(\cdot)$ is the distribution function of X_1. Based on the above discussion, we propose the following VDR algorithm for generating random variate X having a density, $f(x)$:

(1) Generate a pair of uniform random numbers (U, V);

(2) Find a_l and a_u, $(a_l < a_u)$ to satisfying

$$\begin{cases} U = 1 - F(a_u) + F(a_l) + f(a_u)(a_u - a_l) \\ f(a_u) = f(a_l), \qquad a_l < a_u. \end{cases} \tag{4.16}$$

(3) Deliver X, where $X = a_l + (a_u - a_l)V$.

The second step of this algorithm is based on the inverse transformation method for generating the distribution $F_2(v)$ of X_2 and the third step generates the uniform distribution on $S_f(v) = [a_l, a_u]$. In the following, we discuss how to find a_l and a_u in step (2).

We can apply the bisection method to solve equation (4.16) by combining (4.16) and (4.15). More precisely, we have the following steps:

(a) Take $v_l = 0$, $v_u = M$ and $v_m = M/2$; choose an error bound δ;

(b) Find $a_l < a_u$ in Eq. (4.15) with $v = v_m$.

(c) Calculate

$$\Delta = U - [1 - F(a_u) + F(a_l) + f(a_u)(a_u - a_l)].$$

(d) If $|\Delta| < \delta$, deliver a_l and a_u and terminate the process; otherwise go to the next step;

(e) If $\Delta < 0$, take $v_u = v_m$ and $v_m = (v_l + v_u)/2$ and go to Step b; otherwise take $v_l = v_m$ and $v_m = (v_l + v_u)/2$ and go to Step b.

Remark. When the support of $f(\cdot)$ is the finite interval $[l, u]$ and $\max\{f(u), f(l)\} > 0$, the solutions of equation (4.15) are as follows

$$
\begin{aligned}
a_l &= l \quad \text{if} \quad v < f(l), \\
a_u &= u \quad \text{if} \quad v < f(u).
\end{aligned}
$$

If the support of $f(x)$ is $[0, \infty)$ and f is decreasing over $[0, \infty)$, then formula (4.13) becomes

$$F_2(v) = 1 - F(a_u) + f(a_u)a_u \tag{4.17}$$

and steps (b)-(e) are replaced by

(b) Find a_u satisfying

$$U = 1 - F(a_u) + f(a_u)a_u, \qquad f(a_u) = a_u \tag{4.18}$$

(c) Deliver X, where $X = a_u V$.

Similarly, we can treat the case when the support region of $f(x)$ is $(-\infty, 0]$. Alternatively, we can give an approximate algorithm. Denote by $g_1(\cdot)$ the inverse function $f(\cdot)$, on $(a, \infty]$. Let $h_i = (iM/N), i = 0, 1, \ldots, N$, for some positive integer N. Set

$$
\begin{aligned}
p_i &= \int_{h_i}^{M} g_1(y)dy + \int_{h_i}^{M} g_2(y)dy \\
&= p_i + 1 + \int_{h_i}^{h_i+1} g_1(y)dy + \int_{h_i}^{h_i+1} g_2(y)dy, \qquad p_0 = 1.
\end{aligned}
$$

The approximate algorithm is as follows:

(1) Generate (U, V), a pair of uniform random numbers;
(2) Let k be the integer satisfying $h_k < U < h_{k-1}$;
(3) Deliver $x = g_2(h_k) + (g_1(h_k) - g_2(h_k))V$.

When the support region of f is $[0, \infty)$ and $f(0) = M = \max\{f(x) : x \geq 0\}$, the algorithm simplifies to:

$$p_i = \int_{h_i}^{M} g_1(y)dy \quad \text{and} \quad x = g_1(h_k)V.$$

4.7 The Uniform Distribution on a Manifold in $\Re^n$

We shall now briefly consider the uniform distribution on a manifold in $\Re^n$. Let S^n be an $(n-1)$-dimensional manifold in $\Re^n$ and $\mathbf{X} = (X_i, \ldots, X_n)$ be a random vector following the uniform distribution on S_n. Usually S_n is defined as

$$S_n = \{\mathbf{x} : H(\mathbf{x}) = 0, \mathbf{x} \in \Re^n\} \tag{4.19}$$

where $H(\mathbf{x})$ is a function defined on $\Re^n$ with the required properties. Let $L_n(S_n)$ denote the area of S_n. It is well known that (Pearson, 1974).

$$\begin{aligned} L_n(S_n) &= \int_{D_n-1} \frac{\sqrt{(\partial H/\partial x_1)^2 + \cdots + (\partial H/\partial x_n)^2}}{(\partial H/\partial x_n)} dx_1 \cdots dx_{n-1} \\ &\equiv \int_{D_n-1} f_c(\mathbf{x}_{n-1})dx_1 \cdots dx_{n-1}, \end{aligned} \tag{4.20}$$

where $D_{n-1} = \{\mathbf{x}_{n-1} = (x_1, x_2, \cdots, x_{n-1})$. Then, there is an x_n such that $H(\mathbf{x}_{n-1}, x_n) = 0\}$. Without loss of any generality, suppose there is only one such x_n and it is considered as a function of x_{n-1}. Tashiro (1977) and Fang and Wang (1994, Section 4.3) gave a stochastic representation of the uniform distribution on a spherical surface. By lemma (4.1) we could give a general representation of the uniform distribution on a surface as below (Theorem 3 of Fang *et al.*, 2001):

Theorem 4.3 *Let S_n and D_{n-1} be defined as above and let all partial derivatives of H be continuous, then $\mathbf{X}_n = (\mathbf{X}_{n-1}, X_n)$ has the uniform distribution on S_n iff $\mathbf{X}_{n-1}$ has density $f_c(\mathbf{x}_{n-1})\mathbf{I}_{D_{n-1}}(\mathbf{x}_{n-1})$.*

Through Theorems 4.2 and 4.3, we can generate the uniform distribution on a surface .

Example 4.6 (Example 4 of Fang *et al.*, 2001). Generation of the uniform distribution on the surface S_3, where

$$S_3 = \{(x, y, z) : z = e^{x^2+y^2}, \ x^2 + y^2 \le 1\}.$$

By Theorem 4.3, it is equivalent to generate random vectors (X, Y) with density

$$p(x, y) = \frac{1}{c}\sqrt{1 + 4(x^2 + y^2)e^{2(x^2+y^2)}}, \quad x^2 + y^2 \le 1,$$

where $c = L_2(S_3) = \int_{x^2+y^2 \le 1} p(x, y)dxdy$. It is not difficult to see that

$$
\begin{aligned}
D_p(v) \ &= \ \{(x, y) : p(x, y) \ge v, \ x^2 + y^2 \le 1\} \\
&= \ \begin{cases} S_2(1) = \{(x, y) : x^2 + y^2 \le 1\}, & v < \frac{1}{c}, \\ S_2\left(g^{-1}\left(\frac{c^2v^2-1}{2}\right)\right), & \sqrt{1 + 4e^2} \ge v > \frac{1}{c}, \end{cases}
\end{aligned}
\tag{4.21}
$$

where $g(r) = re^r$ and g^{-1} is the inverse function of g. It is easy to see

$$
L_2(D_p(v)) = \begin{cases} 2\pi, & v < \frac{1}{c}, \\ 2\pi g^{-1}\left(\frac{c^2v^2-1}{2}\right), & \sqrt{1 + 4e^2} \ge v \ge \frac{1}{c}. \end{cases}
$$

By Theorem 4.2, we get the following algorithm:

(1) Let U_1 and U_2 be independent uniform random variates on $[0, 1]$.
(2) If $U_1 < (1/c)$, then $X = \cos(2\pi U_2)$ and $Y = \sin(2\pi U_2)$ otherwise $X = r\cos(2\pi U_2)$, $Y = r\sin(2\pi U_2)$ and $r = g^{-1}((c^2U_1^2 - 1)/2)$.
(3) Let $Z = e^{X^2+Y^2}$ and deliver (X, Y, Z), which has the uniform distribution on S_3.

4.8 Comments

In this chapter, we presented an important application of VDR. That is, we used the VDR method to generate univariate and multivariate random variates. The proposed algorithms are new. Some of the algorithms are shown empirically to be computationally more efficient than the existing algorithms.

Chapter 5

VDR and Chaos

Introduction to Chaos

In this chapter we consider the application of some recent results in distribution theory to the probabilistic analysis of the orbits of real chaotic processes. Necessary and sufficient conditions for such orbits to have a uniform distribution are obtained. A new class of chaos functions is proposed. This class has uncountable cardinality. Each member in this class has a uniformly distributed orbit in the unit interval. Each member of this class has zero as its only rational fixed point. We use this class to construct some generators of very long period. Test results suggest that this class of uniform random number generators has exceptionally good auto-correlation and spectral density properties.

The study of mathematical chaos, or chaotic dynamical systems, has attracted much interest in recent years. Comprehensive introductions can be found in, for example, Devaney (1989) and Ott (1993). A number of readable articles offer introductions and are tailored to specific fields or problems. See, for example, Riggle and Madey (1997) for management science and operations research, Feichtinger (1996) for an emphasis on operations research and economics, Bullard and Butler (1993) and Butler (1990) for economics, Crutchfield *et al.* (1982) for physics, McCaffrey *et al.* (1992) for statistics, and Konno and Kondo (1997) for random number generation. Important mathematical properties of chaos on a real interval were obtained by Li and Yorke (1975), which we review below.

To quote Ornstein (1995), "Chaos theory is often introduced by some statement to the effect that dynamical systems that are governed by simple

101

deterministic laws can exhibit behavior that looks random or is essentially random, i.e. small differences in initial conditions grow exponentially, making prediction very difficult, and orbits very complex." Li and Yorke (1975) give precision to these concepts. Such deterministic laws require nonlinearity. The nonlinearity can be quite simple, however, as exemplified by the logistic function and the piece-wise linear tent function, both of which are discussed below.

Our interest is primarily in developing classes of uniform random number generators based on chaos as an application of VDR. Apparently, Konno and Kondo (1997) were the first to propose generation of uniform random numbers using modifications of the tent function and a cubic map. Here, we develop a class of uniform random number generators that are perturbations of tent functions, which we call the *sharkfin* class. We also obtain a general method to plot the density function for chaos processes. Our results are essentially an application of Theorem 1.2.

5.1   The Relationship between VDR and Chaos

Remark on notation: In this chapter, we depart from earlier notation usages. For this chapter, the notation, $f(x)$, and its variants will be used for chaos generator functions. The notation, $\phi(x)$, will denote a probability density function (pdf) with distribution (CDF) function, $\Phi(x)$. The previous use of the notation, $\phi(v)$, is replaced by $\tau(v)$. Thus, in these new notations, the result of Theorem 1.2 may be stated as

$$\phi(x) = \tau(V(x)),$$

where

$$\tau(v) = -\frac{g(v)}{A(v)}.$$

Let $f(x)$ be a continuous function from $[a, b]$ to $[a, b]$. Suppose the recursion $R_0 = \{x_0 : x_{n+1} = f(x_n)\}$ is chaotic. The set of points generated by this recursion is called the *orbit*. We consider the distribution, $\Phi(u)$, of values generated by this recursion, where $\Phi(u) = P\{X_n \leq u$ for all $n \geq 0\}$. In particular, we consider necessary conditions that a uniform density function, $\phi(u)$, exists for such distributions.

We first discuss vertical density representation (VDR) and the vertical density idea. If the orbit values of a chaotic sequence have a probability density function, then the pdf is equal to its vertical density with respect to the generator function, $f(x)$. VDR is therefore a general purpose tool for this class of questions. We also examine its application to a chaotic logistic and a simple piecewise linear $f(x)$. It is shown that the density cannot be uniform in either case. In the case of the logistic function, the pdf must be zero at all points, which are periodic, including single period or fixed points. It is unbounded for at least one point also. In the case of the piecewise linear example, the density cannot be anywhere continuous.

Konno and Kondo (1997) also propose and test uniform random number generators based on chaos functions. They consider recursions for which an explicit solution can be obtained for its sequence value in terms of the term number, n. For example, consider the simple piecewise linear tent function (also called the "baker's transformation" by Konno and Kondo, 1997) given by

$$y_{n+1} = f(y_n) = \begin{cases} 2y_n & 0 \leq y_n \leq \frac{1}{2} \\ 2(1-y_n) & \frac{1}{2} < y_n \leq 1 \end{cases} \tag{5.1}$$

with initial value y_0. The explicit solution

$$y_n = \frac{1}{\pi} \cos^{-1}(\cos 2^n \pi y_0) \tag{5.2}$$

is known. The apparent advantage of such explicit generators is that they cannot be trapped at a fixed point. In contrast, the recursive formulae in (5.1) might terminate at $y_n = 0$ for finite value of n.

To see this, note that if the recursion reaches 0.5 then zero is obtained in two more iterations. In addition, 0.5 has an infinite number of pre-images of the form $0.5(2)^{-n}, n \geq 1$. Thus there is a large chance that the recursion based on (5.1) will terminate using the rational arithmetic of the computer. Clearly, zero is a fixed point and we may describe $0.5(2)^{-n}, n \geq 0$ as *pre-fixed points*. In fact, it is clear that if a generator has any rational fixed point then finite termination is likely for similar reasons. In the generators we propose, nonzero fixed points can often be chosen as irrational or equivalently, *computer irrational*. By that term, we mean numbers that are irrational or numbers such as $\frac{1}{6}$, which cannot be represented with finite precision or obtained by round-off. In that case there is no concern about termination of the process when starting from a nonzero initial point. However, cycles may still occur with length depending on machine precision. Also, provisions for restarting in case of zero round-off need to be given.

It appears that the study of Konno and Kondo (1997) used recursions due to a concern expressed for terminations at fixed points. Further limitations on the practical computing use of the formula-based generators are related to storage and computing time considerations. In the use of (5.2), it would be necessary to store n, the number of random numbers returned since the beginning of the program. Also, the evaluations of the cosine and arccos functions in 5.2 are likely to require considerably more time than use of recursions. Finally, recursions need to store only the most recent return and not the sequence number of this return.

5.2   Recursions and the Vertical Density

Suppose the recursion $\{x_0 : x_{n+1} = f(x_n)\}$ is given for $f(x)$ a continuous function from $[a, b]$ to itself. If the orbit $\{x_n\}$ has a density $\phi(x)$, then since x_{n+1} is exactly the same sequence except for a singleton set, its density must also be $\phi(x)$. To see this, note that we may label each iterate as v_n or x_{n+1}. As will be seen in Theorem 5.1 below, the density $g(v)$, if it exists for $v = f(x)$, must be $\phi(v)$. To apply the foregoing VDR facts, consider

$$S(u) = \{x : f(x) \geq u\}, \tag{5.3}$$

with

$$A(u) = L\{x : f(x) \geq u\} \tag{5.4}$$

For the examples we consider, $A'(u)$ exists piecewise throughout the interval $[a, b]$. We have

Theorem 5.1   *Let $\{x_0 : x_{n+1} = f(x_n), x_n \in [a, b], \text{ for all } n\}$ be a sequence generated by a starting point x_0, where $f(x)$ is a continuous function from $[a, b]$ into itself. Let $A(u) = L\{x : f(x) \geq u\}$ be a differentiable function on $[a, b]$. Then, if a density $\phi(x)$ exists, i.e., $P\{x_n \in A\} = \int_A \phi(x)dx$ for all measurable A, then it must satisfy*

$$\phi(f(x)) = -\phi(x)A'(f(x)) \tag{5.5}$$

for almost all $x \in [a, b]$.

Proof. We note the orbit $\{x_n : n = 0, 1, ...\}$ and the orbit $\{x_{n+1} : x_{n+1} = f(x_n), n = 0, 1, ...\}$ differ at most on a set of measure zero. Let this density be $\phi(x)$. We may label each iterate as v_n or x_{n+1}. That is, $x_{n+1} = f(x_n) = V(x_n) = v_n$. Then, except for x , the same orbit is generated in the support set $[0, 1]$ of the vertical pdf $g(v)$, as in the support set $[0, 1]$ of the $\phi(x)$-pdf. It must be that these have the same densities when they exist. That is, $g(v) = \phi(v)$. Hence, if $\phi(x)$ is the probability density function for one, it is also for the other. Since $x_{n+1} = f(x_n)$, the pdf for x_{n+1} is the vertical density for $\phi(x)$. By Theorem 1.2, in the present notations, and with $f(x) = V(x)$,

$$\phi(x) = \tau(f(x)) = -\phi(f(x))/A'(f(x)) \qquad (5.6)$$

Hence,

$$\phi(f(x)) = -\phi(x)A'(f(x))$$

Theorem 5.2 *Necessary and sufficient conditions that the recursion $\{x_0 : x_{n+1} = f(x_n)\}$ produces uniformly distributed values in $[0, 1]$ are that: (1) the orbit $\{x_n\}$ is dense in $[0, 1]$, and (2) that $A'(f(x)) = -1$ almost everywhere in $[0, 1]$.*

Proof.

(Necessity): If the series values are uniformly distributed then they must clearly be dense in $[0, 1]$, with a density of $f(x) = 1$ almost everywhere. It follows that $f(x_{n+1}) = f(x_n) = 1$ almost everywhere and that $A'(f(x)) = -1$ from (5.5).

(Sufficiency): If $A'(f(x)) = -1$ then $f(x_{n+1}) = f(x_n)$ almost everywhere, so that the density is constant almost everywhere on $[0, 1]$.

5.3   The Logistic Chaos Generator

The logistic function $f(x) = \lambda x(1 - x)$ is known to produce chaotic orbits for $\lambda \in [a_0, 4]$, where a_0 is approximately 3.57. This section examines the implications of the above results for this example. The range of values generated by the recursion $x_{n+1} = 3.89x_n(1 - x_n)$ is easily seen to be $[0.1040332, 0.9725]$.

Ott (1993) obtains the density of orbit values for the logistic map for the special case of $\lambda = 4$. See also Konno and Kondo (1997). This pdf is given by $\phi(x) = \pi^{-1}[x(1-x)]^{-1/2}$, the arcsine density on $(0,1)$. In this section, we propose a graphical procedure for plotting orbit densities that applies to general recursion chaos generators. In particular, our procedure applies for cases of the logistic map when $\lambda < 4$. More generally, this procedure can be used for any chaos generator function for which the derivative, $A'(v)$ can be evaluated. We believe that this direct approach will be of interest as an alternative to histogram or other methods of density estimation. See, for example, Silverman (1986), for a survey of density estimation methods.

Yamaguchi and Hata (1983), see also Konno and Kondo (1997), obtain the explicit solution for the logistic map with $\lambda = 4$ as $x_{n+1} = \sin^2(2^n \sin^{-1} \sqrt{x_0})$. However, in this section, we deal only with the recursion and consider other values of λ for which an explicit solution may not be available. In addition, our method is not limited to the logistic map.

Here, we take $V(x) = f(x)$ as the logistic function and consider $A(v) = A(f(x))$. The function $A(v)$ associated with this example can be obtained as follows. The roots of the equation

$$v = 3.89x(1-x) \tag{5.7}$$

are given by

$$x = \frac{1}{2} \pm \frac{1}{2}\sqrt{\left(1 - \frac{4v}{3.89}\right)}. \tag{5.8}$$

Unlike the case with $\lambda = 4.0$ for which the logistic function has range $[0,1]$, when $\lambda = 3.89$ and it has range $[0.1040332, 0.9725]$. Moreover, at the value of $\lambda = 3.89$, the logistic function at $x = 0.1040332$ is 0.3636295. This requires $A(v)$ to be defined differently on two segments as follows:

$$A(v) = \begin{cases} \sqrt{\left(1 - \frac{4v}{3.89}\right)} & \text{for } 0.362588 \le v \le 0.9725 \\ 0.3959628 + 0.5\sqrt{\left(1 - \frac{4v}{3.89}\right)} & \text{for } 0.1040332 \le v \le 0.362588 \end{cases} \tag{5.9}$$

It follows that

$$A'(v) = \begin{cases} 0.5 \times \dfrac{1}{\sqrt{\left(1 - \frac{4v}{3.89}\right)}} \times \left(\frac{-4}{3.89}\right) & \text{for } 0.362588 \le v \le 0.9725 \\ 0.25 \times \dfrac{1}{\sqrt{\left(1 - \frac{4v}{3.89}\right)}} \times \left(\frac{-4}{3.89}\right) & \text{for } 0.1040332 \le v \le 0.362588. \end{cases} \tag{5.10}$$

Since $A'(v) \neq -1$ for all v, the density of the sequence values cannot be the uniform density. We may note that if a density $\phi(x)$ does exist for these sequence values, then it must satisfy the functional equation (5.5), where $f(x)$ is given by $f(x) = 3.89x(1-x)$, $x \in [0.1040332, 0.9725]$ and $A'(f(x))$ is calculated by way of (5.10). While this functional equation appears difficult to solve analytically, a scalar multiple of the density may be plotted as the chaotic sequence is evolved. Namely, let $\phi(x_0) = c$. Then

$$
\begin{aligned}
\phi(x_1) &= -cA'(x_1), \\
\phi(x_2) &= -\phi(x_1)A'(x_2) = cA'(x_1)A'(x_2),
\end{aligned}
$$

and generally,

$$
\phi(x_n) = (-1)^n c \prod_{k=1}^{n} A'(f(x_k)), \tag{5.11}
$$

where the value of $c = \phi(x_0)$ is that value for which the integral $\int_0^1 \phi(x)dx$ is unity and $x_{n+1} = f(x_n)$.

Thus, using (5.11) we can recursively plot the pdf $\phi(x)$ up to a scalar multiple. We may estimate the integral of the plotted function by numerical integration, for example, as follows. Let us assume that $c = \phi(x_o) = 1$. We may then generate the list $\{(x_{(0)}, \phi(x_{(0)})), (x_{(1)}, \phi(x_{(1)})), \ldots\}$, where the $\{x_{(0)}, x_{(1)}, \ldots\}$ are the $\{x_0, x_1, \ldots\}$ sorted in ascending order. We propose a modification of the trapezoid numerical integration formula to approximate the integral, K_n, after n terms. Unless $x_{(0)} = 0$ and/or $x_{(n)} = 1$, the recursions will not reveal any information on the values of $\phi(0)$ and $\phi(1)$. Therefore, we modify the trapezoid formula by using rectangular approximations for the first and last subintervals. This results in the formula:

$$
\begin{aligned}
K_n &= x_{(0)}\phi(x_{(0)}) + \left\{ \sum_{i=1}^{n} (x_{(i)} - x_{(i-1)}) \left(\frac{1}{2}\phi(x_{(i)}) + \frac{1}{2}\phi(x_{(i-1)}) \right) \right\} \\
&\quad + (1 - x_{(n)})\phi(x_{(n)})
\end{aligned}
$$

Finally, if all the ordinates of the graph are divided by K_n, we have a graph that approximates the true density.

5.4 The Uniform Density

By Theorem 5.2, a chaotic sequence can have a uniform density if and only if $A'(f(x)) = -1$, i.e., $A'(v) = -1$, where $v = f(x)$, the generator function of the sequence. One simple generator function with this property is given by

$$f(x) = \begin{cases} 2x & 0 \leq x \leq 0.5 \\ 2(1-x) & 0.5 \leq x \leq 1.0, \end{cases} \tag{5.12}$$

since $A(v) = 1 - v$, $0 \leq v \leq 1$.

Konno and Kondo (1997) obtain the uniform density result for the generator in (5.2) using the equation, $\rho(x) = \int \rho(x)\delta[x - f(y)]dy$, which they call the *Frobenius-Peron equation*. (Ott, 1993, describes this same result as a limit of $\rho_{n+1}(x) = \int \rho_n(x)\delta[x - f(y)]dy$, also called there by that name.) These authors also note that the logistic map with $\lambda = 4$ can be used with the transformation $y = 2\pi^{-1}\sin^{-1}\sqrt{x}$ to produce a uniform distribution. However, it may be checked that this map has a nonzero rational fixed point, $x = 0.75$, which may limit the usefulness of that approach. Similarly, these authors consider a cubic map, $f(x) = x(3 - 4x)^2$. The reader is referred to Konno and Kondo (1997) for details. Again, we see that this map has two nonzero fixed points at $x = \frac{1}{2}$ and $x = 1$, respectively. Thus, that map may not be of practical interest. Nevertheless, some computational tests given by Konno and Kondo (1997) indicate surprisingly good performance for those classes of uniform generators. However, it is clearly desirable that all non-zero fixed points of a map should be computer irrational. In that case, termination can only occur by round-off to zero, or prior round-off to unity. The generator class we propose permits the choice of parameters with this property.

However, it is clear that iteration of $x_{n+1} = f(x_n)$ with rational computer arithmetic will have finite, and relatively short periods for any rational starting point. Hence we consider what may be called a *strata shift* of this function. First, the inverse functions related to (5.12) are given by

$$\begin{cases} x^-(v) = 0.5v & \text{if } 0 \leq x^-(v) \leq 0.5 \\ x^+(v) = 1 - 0.5v & \text{if } 0.5 \leq x^+(v) \leq 1. \end{cases} \tag{5.13}$$

Consider the modifications given by $x_\alpha^-(v) = 0.5v - \alpha v(1-v)$ and $x_\alpha^+(v) = x_\alpha^-(v) + (1-v) = 1 - 0.5v - \alpha v(1-v)$. Use of the $-\alpha v(1-v)$ perturbation term simplifies resulting formulas as compared with the choice of the

$+\alpha v(1-v)$ form.

Hence, $A(v) = x_{\alpha}^{+}(v) - x_{\alpha}^{-}(v) = 1 - v$. So that $A'(v) = -1$ holds. In order that $0 \le x_{\alpha}^{-}(v) \le 0.5$ and $0.5 \le x_{\alpha}^{+}(v) \le 1$, it is necessary that $-0.5 < \alpha < 0.5$. If $0 < \alpha < 0.5$ then the sharkfin curve bows to the left and bows to the right for the opposite case. If $\alpha = 0$ then we have the original tent function, (5.12), which we believe is not of practical interest for simulation due to the possibility of termination discussed earlier. This point is also noted by Konno and Kondo (1997).

The corresponding generator function $f_{\alpha}(x)$ can be obtained by inverting $x_{\alpha}^{-}(v)$ to $f_{\alpha,1}(x)$ and $x_{\alpha}^{+}(v)$ to $f_{\alpha,2}(x)$, so that

$$f_{\alpha}(x) = \begin{cases} f_{\alpha,1}(x) & 0 \le x \le 0.5 \\ f_{\alpha,2}(x) & 0.5 \le x \le 1. \end{cases} \tag{5.14}$$

These inversions yield

$$f_{\alpha}(x) = \begin{cases} (2\alpha)^{-1}\left(\alpha - 0.5 + \left\{(\alpha - 0.5)^2 + 4\alpha x\right\}^{0.5}\right) & 0 \le x \le 0.5 \\ (2\alpha)^{-1}\left(\alpha + 0.5 - \left\{(\alpha + 0.5)^2 - 4\alpha(1 - x)\right\}^{0.5}\right) & 0.5 \le x \le 1. \end{cases} \tag{5.15}$$

A graph of this function is given in Figure 5.1 in Section 5.4 below. The non-zero fixed point of $f_{\alpha}(x)$ is

$$x^0 = (4\alpha)^{-1}\left[3 + 2\alpha - \left\{4\alpha^2 - 4\alpha + 9\right\}^{0.5}\right], \tag{5.16}$$

which can be expected to be irrational for many values of α, since the bracket expression is not a perfect square. The point, $x = 0$, is the only other fixed point. To further discuss the chaos properties of (5.15), we have

Theorem 5.3 *(Li and Yorke, 1975): let $x = [a, b]$ and $f : X \to X$ be*

continuous. If there is an $x^* \in X$, called a Li-Yorke point, such that

$$f^3(x^*) \le x^* < f(x^*) < f^2(x^*)$$

then

(i) for every integer $n \ge 1$, there is a periodic point $x_n \in X$ with period n

(ii) there is an uncountable set $W \subset N(X)$, the set of nonperiodic points in X, satisfying

(a) *If $x, y \in W$ with $x \neq y$, then*

$$\overline{\lim}_{n \to \infty} \; |\, f^n(x) - f^n(y)| > 0 \; \text{ and } \; \underline{\lim}_{n \to \infty} |\, f^n(x) - f^n(y)| = 0.$$

(b) *If $x \in W$ and $y \in P(X) = W - N(X)$, then*

$$\overline{\lim}_{n \to \infty} \; |\, f^n(x) - f^n(y)| > 0$$

It may be noted that for each $\alpha \in (-0.5, 0.5)$ we have also that $x^*(\alpha) = 0.25(1 - \alpha)$ is a Li-Yorke point. To see this, some elementary algebra shows that $f(x^*(\alpha)) = 0.5$. It follows that $f^2(x^*(\alpha)) = 1$ and $f^3(x^*(a)) = 0$. Therefore, $f^3(x^*(\alpha)) \leq x^*(\alpha) \leq f(x^*(\alpha)) \leq f^2(x^*(\alpha))$ and $x^*(a)$ is a Li-Yorke point for this generator.

Thus, the conditions of Theorem 5.3 are met for each $\alpha \in (0, 0.5)$. Furthermore, $0.5 = f_\alpha^{-1}(1)$, $x^*(\alpha) = f_\alpha^{-1}(0.5)$, and more generally, points of the form $f_\alpha^{-n}(1)$ may be called pre-fixed points, as noted earlier, in the sense that f_α^n applied to such points for sufficiently large values of n will terminate the orbit at $x = 0$. Since all such pre-fixed points are smaller than 0.5, we expect starting points, for example in the interval $(0.5, 1)$, should produce chaotic orbits for every $\alpha \in (0, 0.5)$. It may be checked that for rational starting points, x_0, in $(0.5, 1)$ and for every $\alpha \in (0, 0.5)$, $f(x_0)$ is irrational so that the orbit cannot terminate at 0 in theory. However, due to computer roundoff, provision needs to be made for possible restarts if a termination occurs due to rational approximation. Therefore, if we provide for an arbitrarily large number of restarts, the result is expected to be of indefinite period from a practical computing perspective. We propose and test a procedure which utilizes the logistics generator to produce restart values from $\alpha \in (0.0, 0.5)$.

Uniform Chaos Algorithm: Chaos-Based Uniform Random Generator with Logistic Restarts:
The procedure is as follows:

<u>Initialization:</u>   Choose seeds as follows:

$$\begin{aligned}
&x_0 \in (0.5, 1) && \text{(main process seed)} \\
&\lambda_0 \in (3.54^+, 4) && \text{(logistic restart process seed)}
\end{aligned}$$

$$\alpha_0 \in (0, 0.5) \qquad \text{(alpha seed)}$$

Initialization of running variables:

$$x = x_0$$
$$\alpha = \alpha_0$$

Random Number Call Routine:

If $x \neq 0$ then
$$\widetilde{x} = f_\alpha(x)$$
$$x = \widetilde{x}$$
Return x and exit routine.
Else $(x = 0)$
$$x = x_0$$
$$\widetilde{\alpha} = \tfrac{1}{2}\lambda_0 \alpha(1 - \alpha)$$
$$\alpha = \widetilde{\alpha}$$
$$\widetilde{x} = f_\alpha(x)$$
$$x = \widetilde{x}$$
Return x

The logic for the updating of α is as follows. For the chosen λ_o value, the function, $\lambda_o \alpha(1 - \alpha)$, is a chaos-producing logistic function that will return a value in $[0, 1]$. The factor of $\frac{1}{2}$ will then return a value in the alpha seed range. We conjecture that full parameter ranges would suffice for the above generator. Namely, in the initialization step, we might select $x_o \in (-0.5, 0) \cup (0, 0.5)$ and $\alpha_o \in (-0.5, 0) \cup (0, 0.5)$. In that case, the factor of $\frac{1}{2}$ would not be needed for the α-updating formula, except that provision might be desirable to avoid $\alpha = 0.5$. That case could alternatively be remedied by the restart procedure. In this algorithm, no updating of the logistic parameter, λ_o, has been provided. However, that could also be included for extended cycle length, as for example, by making it some function of x, α and its (their) own previous value(s). Similarly, provision might be made for the possibility of encountering a nonzero fixed point - in case it is uncertain whether $x_o(\alpha)$ is irrational. Some computational experiments have been performed with a simplified version of the above algorithm, which omits the restart feature for zero and nonzero fixed points. These are discussed below. No zero or nonzero fixed point terminations have been observed with the simplified algorithm to date.

5.5   Computations of the Sharkfin Generator

In this section, we report on some initial numerical experiments with the sharkfin generator. The authors thank Prof. Gregory R. Madey for providing the calculations and related figures that were carried out using *Mathematica*. See Wolfram (1991) and Wolfram Research, Inc. (1993). These calculations follow approximately the pattern of analysis suggested by Bowman (1995).

Computational experiments were done with the sharkfin generator (5.15) using $\alpha = 0.49$ and $x_o = 0.2$. A graph of this sharkfin function is shown in Figure 5.1.

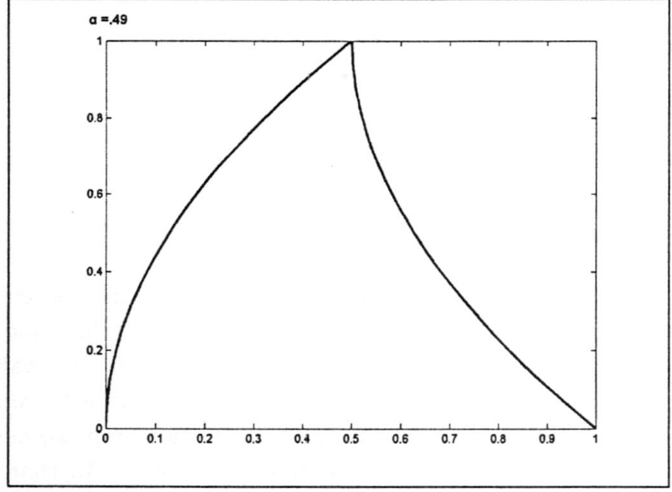

Fig. 5.1. The graph of the sharkfin function for $\alpha = 0.49$

For this value of α, the nonzero fixed point, $x_o(\alpha)$, has significant digits beyond the 16 digit default precision used. ($x_o(0.49) = 0.65399691642056$ 7458^+). It cannot therefore be encountered by means of exact calculation or round-off. These experiments used up to 2048 consecutive iterations to produce the figures. However, in a search for possible termination, $1.0E+12$ iterations have been observed without evidence of termination due to zero or unity round-offs.

No results on cycling have been established to date for this class. A sufficient condition for cycling is that the recursion returns to x_o, either exactly or by round-off. However it is suspected that cycling might occur in other ways or from other points. Moreover, cycle length should be expected to depend heavily on the precision used for the calculations. If a test for a cycle completion can be developed then a restart strategy could be used. As an alternative, a fixed number of recursions, for example, $1.0E + 06$, might be used with restarts.

It is important to note that despite theoretical assurance of the uniform distribution, computer testing is still necessary due to what may be called the *continuous rational restart* property. Assuming that the true iterates of the sharkfin are irrational, then the computed iterates amount to continuously restarting the process from rational round-offs. Thus, we do not expect to be able to follow a true orbit of the process. Also because of the sensitive dependence on initial conditions for chaos generators, the computed results are not even likely to be approximations of the true orbit.

We first consider evidence for the generated distribution actually being uniform. Figure (5.2) shows the one-dimensional plot of the iterates plotted in the order in which they were obtained. Figure (5.3) shows the histogram.

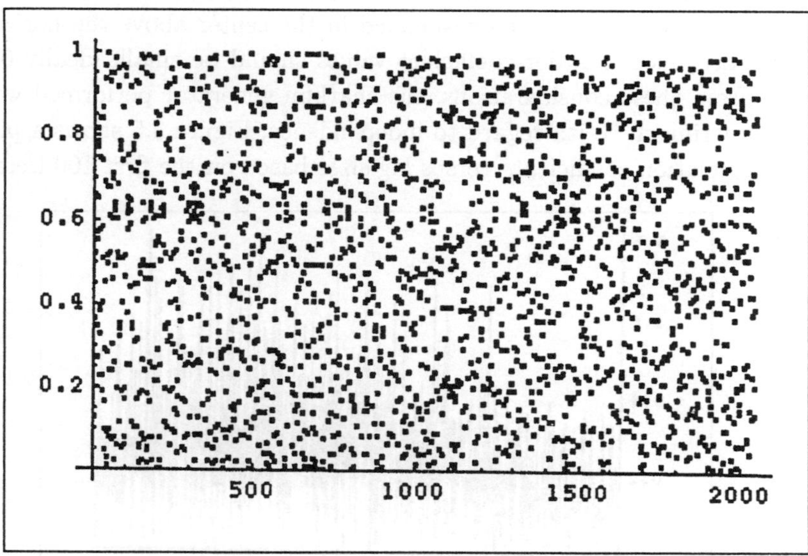

Figure 5.2: The one-dimensional plot

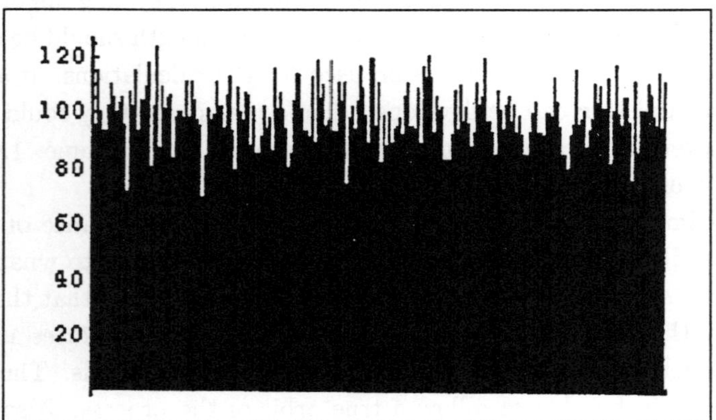

Figure 5.3: The Histogram

Both these are consistent with a uniform density. The observed mean and variance are 0.497399 and 0.0849995, respectively, which compare with the theoretical values of 0.5 and 0.0833333. The associated Chi-square one-sided p-value using ten equally spaced intervals is approximately 0.9982973. The coefficients of the Fast Fourier Transform are shown in Figure 5.4. According to Bowman (1995), for better generators, the dark part of these graphs should appear to be suspended in the center above the horizontal axis. Also, most Fourier coefficient values should be small, ideally 0.6 or less. From these considerations, the sharkfin generator performed well in this experiment. With regard to independence, Figure 5.5 shows a plot of autocorrelation coefficients versus lag sizes based on the first 100 iterates.

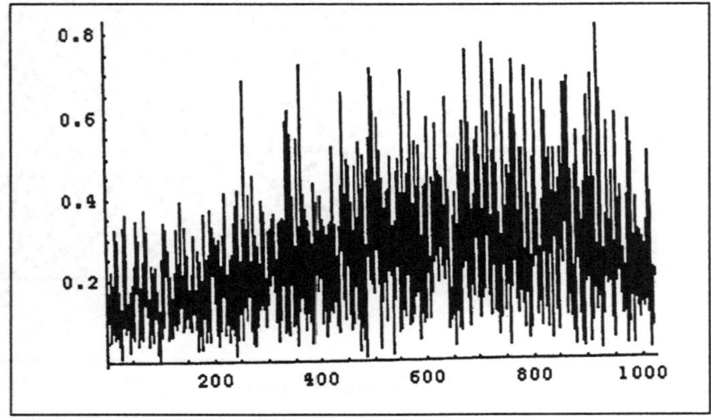

Figure 5.4: The plot of Fast Fourier Transform coefficients

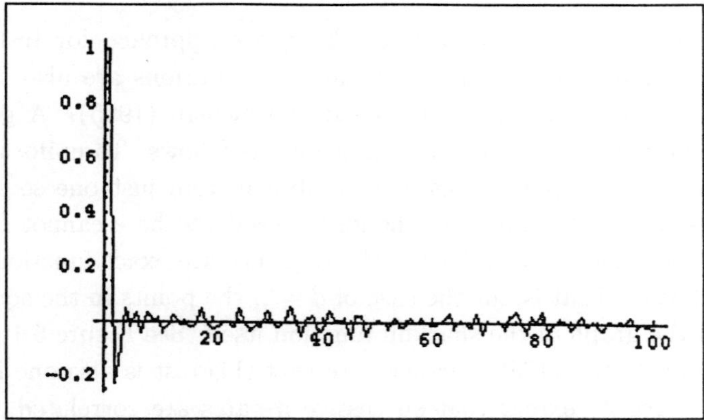

Figure 5.5: The plot of autocorrelation coefficients

Knuth (1981) has suggested that 95% of these coefficients, (for lags greater than zero), should ideally lie in the range from

$$-[2/(N-1)][N(N-3)/(N+1)]1/2 - [1/(N-1)]$$

to

$$[2/(N-1)][N(N-3)/(N+1)]1/2 - [1/(N-1)].$$

For the present case of $N = 100$, Bowman R. (1995) notes that this range is $(-0.21, 0.19)$. Figure 5.5 suggests that the sharkfin appears to conform very well to this criterion. This apparently good performance may seem somewhat counterintuitive in view of the fact that successive iterates are deterministic functions of their predecessors. This aspect can be seen better in the next view of independence for which generators of the exact recursion type cannot perform well.

Law and Kelton (1982) suggest the serial test as another test of uniformity and also independence. If a sequence of U_i's were in fact a sequence of i.i.d. uniform random variables, then the constructed sequences of d-tuples of the form

$$\begin{aligned}
\mathbf{U}_1 &= (U_1, U_2, \ldots, U_d) \\
\mathbf{U}_2 &= (U_{d+1}, U_{d+2}, \ldots, U_{2d}) \\
&\cdots \\
&\cdots
\end{aligned}$$

should be uniform random vectors in the d-dimensional unit hypercube.

Law and Kelton (1982) propose a Chi-square approach for testing this higher dimensional uniformity. Similar considerations are also the basis of the noise sphere approach discussed in Bowman (1995). A generator with this property has practical significance as follows. To uniformly sample the unit hypercube, values may be drawn from just one sequence or one seed stream. Generators of the kind considered here cannot have this property since successive values of the sequence are exact functions of the previous terms. That is, for the case of $d = 2$, the points in the square will consist of the graph of the sharkfin function itself. See Figure 5.1 above.

Law and Kelton (1982) further note that this test is also one for independence. To the extent that successive iterates are correlated, one can expect less uniform sampling of the unit square, for instance. The question arises then as to how to generate uniform distributions on unit hypercubes with the present type of deterministic recursion generators.

Let $U(x_i, \alpha_i)$ be the sequence generated by the sharkfin function with main process seed, x_i, and alpha seed of α_i. A natural approach to generating vectors in the unit d-hypercube is to assign to coordinate i the value from $U(x_i, \alpha_i)$ for $i = 1, \cdots, d$, for each set of iterations. If correlations of these individual dimension sequences are essentially zero, then we may expect uniform coverage of the hypercube. Therefore, such correlation experiments for separate sharkfin generators will be considered in future research. In applications for which multiple independent uniform random number streams are required, it would seem to be good practice to actually obtain correlations of the sequences used as part of the evaluation of the overall experiment. Such sequences should ideally pass a test of being mutually uncorrelated. We also note that even for the popular linear congruential generators, sampling from the unit hypercube can be problematical due to Marsaglia's (1968) linear dependencies observation that has become known as *lattice structure analysis*. Bowman (1995) gives further references on this topic, which is beyond our scope.

5.6 Generalizations of Sharkfin Generators

By Theorem 5.2, any chaos generator, $f(x)$, for which $L[\{x : f(x) \geq v\}] = 1 - v$, will provide a uniformly distributed orbit. It remains for further research to find efficient methods for constructing a wide variety of such generators. Advances in that direction may also enable the use of Theorem

5.1 for designing chaos generators, which have specified orbit densities other than the uniform on $[0, 1]$.

Nevertheless, the simple sharkfin class can be modified in several ways. We have used parametric perturbations of the tent function inverse to construct the sharkfin. These were of the form, $\pm\alpha\psi(v)$, where we chose $\psi(v) = v(1 - v)$. It is merely coincidental that this is also the functional form of the logistic function. In particular, chaos properties of the logistic function have neither been used nor are they required for this function.

Also the construction was based on finding a perturbation function such that $x_\alpha^-(v)$ is monotone increasing on $[0, \frac{1}{2}]$. This enables the simple construction of $x_\alpha^+(v)$ as monotone decreasing on $[\frac{1}{2}, 1]$ by way of setting $x_a^+(v) = x_a^-(v) + 1 - v$. Within this simple setting, the $\psi(v)$-function should ideally have the following properties:

(i) $\psi(0) = \psi(1) = 0$.
(ii) Be bounded so that permissible finite ranges of the α-parameter can be determined.
(iii) Permit analytic inversions required to obtain $f_\alpha(x)$.
(iv) Have irrational or computer irrational nonzero fixed points of $f_\alpha(x)$.
(v) Possess Li-Yorke points for all α in the permissible range.

Thus, a wide variety of functions might be considered for $\psi(v)$. With respect to requirements (i) and (ii), these include all polynomials and bounded rational functions having roots of zero and unity, as well as functions such as $\sin(n\pi x)$, for integers, $n \geq 1$, and their further products and sums.

Also the tent function itself need not be constructed with mode $m = \frac{1}{2}$. Consider the tent function with mode $m, 0 < m < 1$ and perturbation function, $\psi(v)$. We have for $\alpha = 0$

$$f(x) = f_\alpha(x) = \begin{cases} x/m & \text{for} \quad x \leq m \\ \frac{(1-x)}{(1-m)} & \text{for } x \geq m. \end{cases} \tag{5.17}$$

For the modified inverse function on the interval, $0 \leq x \leq m$, we have

$$f_{1,\alpha}^{-1}(v) = x_\alpha^-(v) = mv + \alpha\psi(v).$$

And similarly on the interval, $m \leq x \leq 1$,

$$f_{2,\alpha}^{-1}(v) = x_\alpha^+(v) = mv + 1 - v + \alpha\psi(v)$$

where we have used the $+\alpha\psi(v)$ form of the perturbation term. These functions, $f_{1,\alpha}^{-1}(v)$ and $f_{2,\alpha}^{-1}(v)$, must be inverted to construct $f_\alpha(x)$.

The choices of $m = 1/3$ or $m = 1/6$ are attractive since these are computer irrational. For $m = 1/3$, $f(x) = 3x$ for $x \leq 1/3$ and $f(x) = (3/2)(1 - x)$ for $x \geq 1/3$. However, in that case the nonzero fixed point in $(0, 1)$ is $x = 3/5$, which is rational and likely to lead to termination for the case of $\alpha = 0$. If $m = 1/6$ then $f(x) = 6x$ if $x \leq 1/6$ and $f(x) = (6/5)(1-x)$ if $x \geq 1/6$. Its nonzero fixed point is $x = 6/11$, a repeating decimal, which is computer irrational, being unrepresentable for any finite decimal precision. For illustration, we let us choose $m = 1/6$. Thus, with $\psi(v) = v(1-v)$ and perturbation form, $+\alpha\psi(v)$, these choices yield the

Sharkfin with $m = 1/6$ and perturbation form, $+\alpha v(1 - v)$:

If $\alpha \neq 0$, then:

$$f_\alpha(x) = \begin{cases} (12\alpha)^{-1}\{1 + 6\alpha - [(6\alpha + 1)^2 - 144\alpha x]^{0.5}\} & \text{for } x \leq 1/6 \\ (12\alpha)^{-1}\{-5 + 6\alpha - [(6\alpha - 5)^2 + 144\alpha(1 - x)]^{0.5}\} & \text{for } x \geq 1/6. \end{cases}$$

If $\alpha = 0$, then $f_\alpha(x)$ is given by (5.17) above, which is the limit case of the foregoing formulae.

To find the permissible α-range, we first require that $x_\alpha^-(v) = f_{1,\alpha}^{-1}(v) = mv + \alpha v(1 - v)$ be monotone increasing on $(0, m)$. Similarly, $x_\alpha^+(v) = f_{2,\alpha}^{-1}(v) = mv + 1 - v + \alpha\psi(v)$ should be monotone decreasing on $(m, 1)$. For the present example, we may reason as follows. Looking at the interval $x \in [0, 1/6]$ and $\alpha > 0$, we have $x(v) = x_\alpha^-(v) = (1/6)v + \alpha v(1 - v)$ and $x'(v) = 1/6 + \alpha - 2\alpha v$. Also, we have $x(v)$ concave here. Evidently, $x'(0) > 0$ and $x'(1) = 1/6 - \alpha > 0$ provided $\alpha < 1/6$. Now considering the interval $x \in [1/6, 1]$ and $a < 0$, we have $x(v) = x_\alpha^+(v) = (1/6)v + 1 - v + \alpha v(1 - v)$ and $x'(v) = -5/6 + \alpha - 2\alpha v$. Here, $x(v)$ is concave. In this case, $x'(0) < 0$ and $x'(1) = -5/6 - \alpha < 0$ provided $a > -5/6$. Thus we obtain the permissible a-range as $(-5/6, 1/6)$. Li-Yorke points can be obtained in a manner similar to the case of $m = \frac{1}{2}$ and we omit discussion. The nonzero fixed point for this generator is given by

$$x_o(a) = (12\alpha)^{-1}[-11 + 6\alpha + \{121 + 12\alpha + 36\alpha^2\}^{0.5}]$$

Of course, one might also consider perturbations of the form, $\pm\alpha\sin(n\pi v)$. In fact, $\sin(\pi v)$ has a shape similar to $v(1 - v)$ on $[0, 1]$. Conceptually, this type of perturbation may be just as good except that the inversion steps require solving equations of the form, $a^*v + b^*\sin(n\pi v) = x$. This would likely require a numerical algorithm such as the Newton-Raphson method (see, for example, Kellison, 1975) and may be slower than the square root operation.

5.7 Comments

Using the basic VDR Theorem 1.1, we have been able to establish a design property, $A'(f(x)) = -1$, that a chaos generator $f(x)$ should have in order to produce uniformly distributed orbits in the unit interval. This result also enables a new approach to graphing the orbit densities for chaos generators. Exact numerical use of such generators is not possible due to finite computer arithmetic. Nevertheless, some preliminary computational experiments are encouraging. We believe this is interesting in view of what we call continuous restarts and the sensitive dependence on initial conditions. Further work is needed for understanding the cycle properties for this class. Also, it is hopeful that this class will be able to avoid lattice structure due to the nonlinear dynamics.

The continuous restarts property of these generators, while seemingly presenting a problem, may also offer an opportunity for a more sophisticated class of chaos-based uniform pseudo-random number generators. Namely, using the basic sharkfin function, we might select an α and a starting point, x_o. Then an intermediate x-value, x_o', could be computed by recursion. However, instead of outputting this value, it could be used to generate a revised α-value by way of $\alpha = 0.5x_o'$, for instance. Then using the new α-value and x_o', we may compute x_1 by recursion and output it as the result of the step. This would amount to more dramatically different restarts than those arising merely from round-offs. Such a process would use just the first iterate of each possible chaotic sequence defined by the α and x-values that are encountered. Hopefully, they may still produce orbits, which are acceptably close to being uniformly distributed, while at the same time giving the appearance that outputs are not a deterministic function of the inputs.

Finally, we note that chaos phenomena somewhat blur the distinction between randomness and determinism. A further study of the use of chaos in this topic area may lead to a better understanding of the full set of desirable criteria that pseudo-random numbers should possess.

Chapter 6

Management Science Applications of VDR–I

Introduction

In this chapter, we discuss four topics that apply VDR and are generally related to Management Science topics. The first of these is a topic in the statistics-optimization interface area that we call *Tolstoy's Law of the Mode* or simply, *Tolstoy's Law*. Certain common observations suggest that top performers tend to be similar in many respects. A famous quote of Tolstoy embodies the idea. Section 6.1 proposes an operationalization of the concept and applies it to the relationship between consensus of estimators and accuracy of estimates. A result from this section is used later in Chapter 7 for the development of a validation technique for a new estimation principle.

In Section 6.2, we discuss what we call *normal-like-or-better* performance on the unit interval. In efficiency analysis, observed efficiency scores are distributed on the unit interval. Here, we use VDR to generalize the normal law for this outcome space. Maximum entropy approaches can also be used and we compare the VDR results.

Tolstoy's Law is related to unimodality, which may also be considered an optimization topic. In Section 6.3, some further issues related to unimodality are discussed. For unimodal densities $f(x)$ on the half-interval, we have the associated densities $g(v)$ and $A(v)$ from VDR. In addition, the inverse functions for each of these is a pdf as well. Some relationships among these densities are explicated. We also discuss the relationship of these results to Khintchine's Unimodality Theorem and obtain some results on what we call *Khintchine density representation* (KDR). Finally, we discuss further,

121

what we called strata shift densities in Chapters 2 and 5.

Section 6.4 presents an application of the General VDR Theorem 1.5 to a problem we call *inverse linear programming*. This problem arises when data such as technological constraint coefficients are missing and need to be estimated from observed inputs and outputs of a process. Here, VDR facilitates the construction of a pdf model for use in a maximum likelihood strategy. The VDR construction is also compared to an elementary approach.

6.1 Tolstoy's Law of the Mode (TLM)

Certain common observations suggest that high performers are more alike in some sense than are low performers. A class of functions on $\Re^n$, called Tolstoy functions, is defined and characterized. A statistical converse of is also discussed in terms of what may be called *weak consensus*. Some sufficient conditions for the equivalence of weak consensus and accuracy are also derived.

A well known quote of Tolstoy is as follows (Bartlett, 1992):

> *"Happy families are all alike:*
> *every unhappy family is unhappy in its own way."*

> *Leo Nikolaevich Tolstoy (1828-1910),*
> *Anna Karenina (Part I, Chapter 1, 1873-1876)*

We call the most general form of this observation *Tolstoy's Law of the Mode*. In this section, we propose a quantitative characterization of these observations by defining a Tolstoy's property for functions on $\Re^n$ in terms of performance measurement functions and performance score distributions, namely, the $V(\mathbf{x})$ and $g(v)$ concepts, respectively, from VDR. Using the proposed characterization, we give some sufficient conditions for its validity.

Two observations similar to Tolstoy's quote are:

> *There are more ways to be wrong than to be right.*

and

> *High performers are more alike than are low performers.*

Both validity (wrong versus right) and high performance are analogous to happiness. When TLM holds, high performers in terms of happiness tend to be near to each other in some sense. Low performers in the sense of accuracy (being "wrong") tend to be further apart in some similar sense.

It therefore appears that a very general form of the TLM assertion can be stated as:

> Within a population, members that score highly on a certain criterion will tend to be closer to each other than those who score poorly on the same criterion.

Let a population consist of members possessing feature vectors,

$$\mathbf{x} = (x_1, \ldots, x_n) \in \Re^n.$$

Let $V(\mathbf{x})$ be the criterion function of interest on $\Re^n$. Thus, for example, if the population consists of all families, it is assumed that $V(\mathbf{x})$ represents the happiness level of a family with feature vector, $\mathbf{x}$. For simplicity, it is assumed that $\mathbf{x}$ describes all attributes related to family happiness. Let $\|\mathbf{x}\|$ denote the usual Euclidean norm on $\Re^n$ and let $d(\mathbf{x}, \mathbf{y})$ denote the corresponding metric. Also we write $V(\mathbf{x}) \uparrow v^*$ to mean that the sequence, $V(\mathbf{x}_n)$, is monotonic increasing with limit v^*.

The assertion of TLM can be operationalized as follows. Let v^* be the supremum (possibly $+\infty$) of $V(\mathbf{x})$ on $\Re^n$. Let $\{\mathbf{x}_n\}$ and $\{\mathbf{y}_n\}$ be any two distinct sequences in $\Re^n$ such that $V(\mathbf{x}_n) \uparrow v^*$ and $V(\mathbf{y}_n) \uparrow v^*$. Here, the symbol $\uparrow$ denotes weak monotone convergence. Then for large values of n, both $V(\mathbf{x}_n)$ and $V(\mathbf{y}_n)$ may be said to be "large" as well. The assertion of TLM in this case is that $d(\mathbf{x}_n, \mathbf{y}_n)$ should be "small". That is, $d(\mathbf{x}_n, \mathbf{y}_n) \to 0$ for any two such sequences. Thus we may consider the following specific form of the property.

Definition 6.1: A real-valued function $V(\mathbf{x})$ on $\Re^n$ with $V^* = \sup\limits_{\mathbf{x} \in \Re^n} V(\mathbf{x})$, is said to be a Tolstoy function or to have Tolstoy's property, if for every pair of two distinct sequences $\{\mathbf{x}_n\}$ and $\{\mathbf{y}_n\}$ such that $V(\mathbf{x}_n) \uparrow V^*$ and $V(\mathbf{y}_n) \uparrow V^*$, we have also that $d(\mathbf{x}_n, \mathbf{y}_n) \to 0$.

Remark: It is noted that both $\{\mathbf{x}_n\}$ and $\{\mathbf{y}_n\}$ necessarily converge to the same limit point $\mathbf{x}^*$. Furthermore, if V is defined at this point then $V(\mathbf{x}^*) = V^*$. In this case, $\mathbf{x}^*$ can be called the mode of $V(\mathbf{x})$. Tolstoy's quote can be interpreted in light of these results as the assertion that family happiness is a function with the Tolstoy property.

The first result is as follows.

Theorem 6.1 *Let v^* be the global supremum of $V(\mathbf{x})$. A necessary and sufficient condition that Tolstoy's Law of the Mode holds for the function $V(\mathbf{x})$ on $\Re^n$ is that whenever V_n is a weakly monotone increasing sequence with $V_n \uparrow V^*$, then $D_n \to 0$, where D_n is the diameter of the level set, $\{\mathbf{x} : V(x) \geq V_n\}$.*

Proof: (Necessity). Let $\{V_n\}$ be a sequence such that $V_n \uparrow V^*$. We seek to show that if $V(\mathbf{x})$ has the Tolstoy property, then $D_n \to 0$ where $D_n = diam \{\mathbf{x} : V(\mathbf{x}) \geq V_n\}$. Suppose to the contrary that D_n did not converge to zero. Then for some $\epsilon > 0$, there are points $\mathbf{x}_n$ and $\mathbf{y}_n$ in D_n such that $d(\mathbf{x}_n, \mathbf{y}_n) \geq \epsilon$ for all n. However, $V^* \geq V(\mathbf{x}_n) \geq V_n$ and $V^* \geq V(\mathbf{y}_n) \geq V_n$ imply $V(\mathbf{x}_n) \uparrow V^*$, $V(y_n) \uparrow V^*$. These together with $d(\mathbf{x}_n, \mathbf{y}_n) \not\to 0$ lead to a contradiction to the Tolstoy property. Hence, the necessity.

(Sufficiency) Let $\{\mathbf{x}_n\}, \{\mathbf{y}_n\}$ be two sequences satisfying $V(\mathbf{x}_n) \uparrow V^*$ and $V(\mathbf{y}_n) \uparrow V^*$. Let $\bar{V}_n = \min\{V(\mathbf{x}_n), V(\mathbf{y}_n)\}$. Evidently, $\bar{V}_n \uparrow V^*$. By assumption, $D_n \to 0$ where D_n is the diameter of $S(\bar{V}_n) = \{\mathbf{x} : V(\mathbf{x}) \geq \bar{V}_n\}$. Since $\mathbf{x}_n$ and $\mathbf{y}_n \in S(\bar{V}_n)$, it follows that $D(\mathbf{x}_n, \mathbf{y}_n) \leq D_n \to 0$. This completes the proof of the theorem.

Theorem 6.2 *All of the following functions have the Tolstoy property:*

(a) Unimodal $V(\mathbf{x})$ on $\Re$.
(b) Strictly quasi-concave $V(\mathbf{x})$ on $\Re^n$ provided that a unique maximum point exists.
(c) Strictly monotone increasing $V(\mathbf{x})$ on intervals of $\Re^1$.
(d) Strictly concave functions on closed convex subsets of $\Re^n$.

Proof: In each case the level sets $S(v) = \{\mathbf{x} : V(\mathbf{x}) \geq v\}$ have the property that $S(V_1) \subset S(V_2)$ if $V_1 > V_2$. Therefore the diameters of these sets are nonincreasing. In (a), (c), and (d) the sets $\{\mathbf{x} : V(\mathbf{x}) = V^*\}$ are singletons and have diameter zero. In (b), the uniqueness of the maximum point is enforced for that purpose.

Thus concave functions and their generalizations tend to have the Tolstoy property. However, concavity is not essential. For example, functions with star-shaped level sets and unique modes clearly have the property. Interestingly, Tolstoy functions need not be unimodal. Consider a function $V(\mathbf{x})$ on $\Re$ for which the global maximum point is unique. Sequences $\{x_n\}$ and $\{y_n\}$ for which $\{V(x_n)\}$ and $\{V(y_n)\}$ converge to the global maximum value would eventually converge together. However, if global maximum points are multiple and isolated, this property would not hold for all such sequences. Clearly each sequence could, in that case, converge to a different global maximum point. That is, a Tolstoy function may have other relative maxima so long as it has a unique global maximum or supremum.

As a further example, a constant function cannot have the Tolstoy property since there are no sequences $\{\mathbf{x}_n\}$ for which $V(\mathbf{x}_n)$ is weakly increasing. Similarly, a function that is level at its global maximum could not be a Tolstoy function under Definition 6.1. In view of possible unboundedness at the mode, Tolstoy functions need not be continuous. Also, even in the bounded case, jump discontinuities should not hamper the applicability of Theorem 6.1. However, Tolstoy functions might be said to be *inverse continuous* at $\mathbf{x}^*$, since $V(\mathbf{x}_n) \uparrow V^*$ implies that $\mathbf{x}_n \to \mathbf{x}^*$. In fact, we have

Theorem 6.3 *A real valued function, $V(\mathbf{x})$, on $\Re^n$ has the Tolstoy property iff*

(i) it has a unique global maximum point, $\mathbf{x}^*$, say; and
(ii) it is inverse continuous at $\mathbf{x}^*$.

Proof: (Necessity). As noted above, a Tolstoy function cannot have multiple global maximum points. Hence, (i) holds. For (ii) let $V(\mathbf{x}_n) \uparrow V^*$. We have that $\mathbf{x}_n \in S(V(\mathbf{x}_n))$ and $\mathbf{x}^* \in S(V(\mathbf{x}_n))$ for all n. By Theorem 6.2, $d(\mathbf{x}_n, \mathbf{x}^*) \le D_n \to 0$.
(Sufficiency). $V(\mathbf{x}_n) \uparrow \mathbf{x}^*$ and $V(\mathbf{y}_n) \uparrow \mathbf{y}^*$. By inverse continuity, $\mathbf{x}_n \to \mathbf{x}^*$, $\mathbf{y}_n \to \mathbf{y}^*$ and whence $d(\mathbf{x}_n, \mathbf{y}_n) \to 0$.
In the next section, a kind of converse of Tolstoy's Law is discussed.

6.1.1 *A Stochastic Converse Formulation*

A direct converse of TLM would be similar to the claim that if two population members are near to each other, then both score highly on the criterion function, $V(\mathbf{x})$. This cannot be true, in general, since two points can be arbitrarily close to each other with both having low values of $V(\mathbf{x})$. We therefore consider statistical versions of this claim in which "nearness" of two or more sample points is regarded as indicative of nearness in the probability sense.

The triangle inequality for metrics on $\Re^n$ states that whenever two points $\mathbf{x}$ and $\mathbf{y}$ are close to a third point $\mathbf{z}$, then they are close to each other. That is, if $\|\mathbf{x} - \mathbf{z}\| \leq \frac{1}{2}\epsilon$ and $\|\mathbf{y} - \mathbf{z}\| \leq \frac{1}{2}\epsilon$, then $\|\mathbf{x} - \mathbf{y}\| \leq \epsilon$. The thrust of the present inquiry is to find out when, in a probabilistic sense, a kind of converse of the triangle inequality is true. Namely, under what condition does the closeness of $\mathbf{x}$ and $\mathbf{y}$ imply higher probability of their closeness to $\mathbf{x}^*$, where $\mathbf{x}^*$ is the mode of $V(\mathbf{x})$.

In a sense to be made more precise below, the goal is to obtain conditions under which the relative nearness of two random population members is associated with a high probability of their nearness to the mode. An example which illustrates this idea is as follows. This example uses a Bayesian approach to demonstrate that nearness of expert estimates may be associated with nearness to $\mathbf{x}^*$ in a two dimensional case.

Example 6.1 Let $x \in \Re^2$, $r(\mathbf{x}) = \|\mathbf{x}\|$ and $V(\mathbf{x}) = 1 - r(\mathbf{x})$ for $0 \leq r = r(\mathbf{x}) \leq 1$, and 0 otherwise. Assume $g(v)$ is the uniform density on $[0,1]$. We assume further that two experts attempt to estimate the ideal point $\mathbf{x}^* = (0,0)$ and that $v = V(\mathbf{x}) = \| \mathbf{x} \| = \| \mathbf{x} - \mathbf{0} \|$ is the measure of performance associated with $g(v)$. Consider the case in which two expert estimates $\mathbf{x}^1$ and $\mathbf{x}^2$ of $\mathbf{x}^*$ have identical values of v, and hence identical r-scores. We will first derive the conditional density of r given δ, the distance between $\mathbf{x}^1$ and $\mathbf{x}^2$, and then show that its mode tends to $\mathbf{0}$ (i.e. $\mathbf{x}^1$ and $\mathbf{x}^2 \to \mathbf{x}^*$) as $\delta \to 0$. Hence, this analysis will show that the closer are $\mathbf{x}^1$ and $\mathbf{x}^2$, the more likely they are close to $\mathbf{x}^*$.

First, the reader may check that under the uniform assumption, we obtain for the conditional density of δ given r,

$$h(\delta|r) = \begin{cases} \frac{2}{\pi}(4r^2 - \delta^2)^{-1/2}, & 0 \leq \delta \leq 2r \\ 0, & \text{elsewhere} \end{cases} \tag{6.1}$$

since δ can be at most one diameter. Here, $g(v)$ can also be expressed as a function, $k(r)$, in r which is uniformly distributed on $0 \le r \le 1$. Thus,

$$f(r|\delta) = \frac{h(\delta|r)k(r)}{\int_{\frac{\delta}{2}}^{1} h(\delta|r)k(r)\ dr}, \tag{6.2}$$

noting that for a given δ, r must lie in $[\frac{\delta}{2}, 1]$. Therefore,

$$f(r|\delta) = \frac{\frac{2}{\pi}(4r^2 - \delta^2)^{-1/2}}{\int_{\frac{\delta}{2}}^{1} \frac{2}{\pi}(4r^2 - \delta^2)^{-1/2}\ dr} \tag{6.3}$$

$$= \begin{cases} \frac{4}{\pi}(4r^2 - \delta^2)^{-1/2}(\ln 2 - \ln(2 + (4 - \delta^2)^{1/2}))^{-1}, & 0 \le r \le 1 \\ 0, & \text{elsewhere.} \end{cases} \tag{6.4}$$

It is clear that this pdf has mode $r = \frac{\delta}{2}$. Hence, as δ tends to zero, it follows that the modal value of r, namely $\frac{\delta}{2}$, tends to zero as well, which was to be demonstrated. This example will be discussed further below.

Next, let us consider a random sample $x_1, \ldots, x_m$ from a population in $\Re$ with pdf $f(\mathbf{x})$. Points in this sample are expected to be relatively nearer to each other if they are near to the mode of the population density, $f(\mathbf{x})$, thus relative nearness of a pair of sample points will be associated with their nearness to the mode of $V(\mathbf{x})$, if in particular, the modes of $V(\mathbf{x})$ and $f(\mathbf{x})$ coincide. In the following, we give some sufficient conditions for this coincidence.

It will be useful to first recall some earlier results on conditional contour densities and VDR. Let $V(\mathbf{x})$ be a function on $\Re^n$ with range $[0, V_{\max}]$. Consider the level sets $S(u) = \{\mathbf{x} : V(\mathbf{x}) \ge u\}$. Let $h(\mathbf{x}|u)$ be a family of densities on $\partial S(u) = \{\mathbf{x} : V(\mathbf{x}) = u\}$ of S, the boundary. For a specific value of $u, h(\mathbf{x}|u)$ may be called a contour density.

Let $g(u)$ be a density on $[0, V_{\max}]$, called the vertical, or VDR density. Assume that $\bigcup_{0 \le u \le u_{max}} \partial L_u = \Re^n$. Then Monte Carlo sampling from $g(u)$, followed by similar sampling from the corresponding $h(\mathbf{x}|u)$ will produce variates on $\Re^n$ with a density, $f(\mathbf{x})$, say. Conversely, Monte Carlo sampling from a specified $f(\mathbf{x})$ on $\Re^n$, determines a density of form, $g(v)$, and a density $h(\mathbf{x}|v)$ on the ∂S-sets. Vertical density representation with respect to a specified $V(\mathbf{x})$-function amounts to factoring a density $f(\mathbf{x})$ on $\Re^n$ into vertical (or ordinate) and contour (level set) components.

Let $L(\cdot)$ denote the Lebesgue measure on $\Re^n$. Define $A(u) = L(S(u))$. Since $S(u + \epsilon) \leq S(u)$ for all u, and $\epsilon > 0$, it follows that $A'(u) \leq 0$, in general, when $A(u)$ is differentiable. We consider the special subclass of Tolstoy functions $V(\mathbf{x})$ for which $A'(u) < 0$ for $u \in (0, v_{\max})$.

We further consider the case in which $h(\mathbf{x}|u)$ is the uniform density on the boundary of $S(u)$ for all u. This assumption states that population members which have the same criterion score are uniformly distributed over the corresponding level set of $V(\mathbf{x})$. Recall from Thereom 1.2 that $f(\mathbf{x}) = \phi(V(\mathbf{x}))$ where $\phi(u) = -\frac{g(u)}{A'(u)}$.

Evidently, a sufficient condition that the modes of $V(\mathbf{x})$ and $f(\mathbf{x})$ coincide is that $\phi(u)$ be monotone increasing on $[0, v_{\max}]$. In this case relatively larger values of $f(\mathbf{x})$ are associated with relatively larger values of $V(\mathbf{x})$ and vice versa. In that case, interpoint distances of samples of points from X are smallest near the common mode of $f(\mathbf{x})$ and $V(\mathbf{x})$. Hence, nearness of sample points is thus associated with larger values of $V(\mathbf{x})$.

If $A''(u)$ and $\phi'(u)$ exist, this sufficiency condition can be further analyzed.

Theorem 6.4 *A sufficient condition that $\phi(u)$ is monotone increasing is that*

$$\frac{g'(u)}{g(u)} - \frac{A''(u)}{A'(u)} > 0 \ \text{ for } \ u \in (0, v_{\max}) \ . \tag{6.5}$$

Proof: Taking the logarithmic derivative of $-g(u) = \phi(u)A'(u)$ shows that

$$\frac{g'(u)}{g(u)} = \frac{\phi'(u)}{\phi(u)} + \frac{A''(u)}{A'(u)} \ . \tag{6.6}$$

Therefore

$$\phi'(u) = \phi(u) \left\{ \frac{g'(u)}{g(u)} - \frac{A''(u)}{A'(u)} \right\} \ . \tag{6.7}$$

Thus, for positive $\phi(u)$ we have $\phi'(u) > 0$ provided that $\dfrac{g'(u)}{g(u)} - \dfrac{A''(u)}{A'(u)} > 0$.

Remark: Intuition suggests that strictly increasing monotonicity of $g(u)$ $(g'(u) > 0)$ might be sufficient for this condition. However, the above result shows that geometrical considerations also play a role through the terms related to $A(u)$. The following examples illustrate this.

Example 6.2 The analysis of Example 6.1 above showed that the nearness of points of equivalent $V(\mathbf{x})$-scores was associated with nearness to the mode of $V(\mathbf{x})$. Theorem 1.2 relaxes the requirement that the points in question have equivalent $V(\mathbf{x})$-scores by permitting the broader result that the $f(\mathbf{x})$-pdf has the same mode as does $V(\mathbf{x})$. This can be seen more directly in view of Theorem 6.4. For this example, $V(\mathbf{x}) = 1 - \|\mathbf{x}\|$ for $\|\mathbf{x}\| \leq 1$ so that the mode of $V(\mathbf{x})$ is zero. Here $A(u) = \pi(1 - u)^2$ so that $A'(u) = -2\pi(1 - u)$. Also $g(u) = 1$ and $h(\mathbf{x}|u)$ is uniform for all u. It follows from Theorem 1.2 that

$$f(\mathbf{x}) = \phi(V(\mathbf{x})), \tag{6.8}$$

where

$$\begin{aligned} \phi(u) &= -g(u)/A'(u) \\ &= (2\pi(1 - u))^{-1}, \end{aligned} \tag{6.9}$$

and therefore,

$$f(\mathbf{x}) = (2\pi\|\mathbf{x}\|)^{-1}, \quad \|\mathbf{x}\| \leq 1. \tag{6.10}$$

Thus the mode of $f(\mathbf{x})$ is clearly zero. Furthermore, Theorem 6.4 confirms this result since

$$\frac{g'(u)}{g(u)} - \frac{A''(u)}{A'(u)} = (1 - u)^{-1} > 0 \text{ on } (0, 1) \quad . \tag{6.11}$$

This example offers the following insight. Since the performance pdf is only uniform, the modal concentration at zero is largely due to the graph of $V(\mathbf{x})$. In particular, it has a sharp peak at its mode. If this $V(\mathbf{x})$ is the criterion function for a decision $\mathbf{x}$ in $\|\mathbf{x}\| \leq 1$, then even random performance in the form of uniform $g(u)$ will yield good results. Such a criterion function fails to exhibit the condition sometimes called *flat laxity* (Isaacs, 1965) in which points near to the optimal one are nearly as good as the optimum itself. The same observation is frequently mentioned in discussions of the Economic Order Quantity model in inventory theory (see, for example, Stevenson, 1982). For this $V(\mathbf{x})$, the optimum may be said to be very much better than points nearby. That is, the optimum may be said to be more "obvious".

Example 6.3   Let $\mathbf{x} \in \Re^2$. We may think of $\mathbf{x}$ as the location of a dart thrown at the bull's eye located at $(0,0)$. Let $V(\mathbf{x}) = e^{-\|\mathbf{x}\|^2}$ and $g(u) = 1$ on $[0,1]$. Here $A(u) = -\pi \ln u$, $A'(u) = -\pi u^{-1}$, and $A''(u) = \pi u^{-2}$. It follows that on $(0,1)$

$$\frac{g'(u)}{g(u)} - \frac{A''(u)}{A'(u)} = 0 - \frac{\pi u^{-2}}{-\pi u^{-1}} = u^{-1} > 0 \ . \tag{6.12}$$

Therefore, the mode of the resulting $f(\mathbf{x})$-distribution will be zero. In fact, the reader may check that $f(\mathbf{x})$ is the standard uncorrelated bivariate normal distribution in this case.

Example 6.4   Next consider the analog of Example 6.3 in $\Re$. Namely, $V(x) = e^{-x^2}$, and $g(u) = 1$ on $[0,1]$. For this example, $A(u) = 2(-\ln u)^{1/2}$. Here $A'(u) = -u^{-1}(-\ln u)^{-1/2}$ and

$$A''(u) = -1/2u^{-2}(-\ln u)^{-3/2} + u^{-2}(-\ln u)^{-1/2}.$$

It follows that

$$\begin{aligned}
\frac{g'(u)}{g(u)} - \frac{A''(u)}{A'(u)} &= \frac{1}{2}u^{-1}(-\ln u)^{-1} - u^{-1} \\
&= u^{-1}\left[\frac{1}{2}(-\ln u)^{-1} - 1\right].
\end{aligned} \tag{6.13}$$

However, this value fails to exceed zero consisently in $(0,1)$, for example at $u_0 = \exp(-0.9)$. Thus, by using essentially the same $g(u)$ and $V(x)$, the result in $\Re$ is distinctly different from that in $\Re^2$. This illustrates the impact of the geometric influence through the terms related to $A(u)$.

In order to discuss further examples, consider what may be called *reverse Tolstoy* functions. These are increasing as $\mathbf{x}$ moves away from a point $\mathbf{x}^*$, which may be called the *antimode*. Evidently, a function $V(\mathbf{x})$ is a reverse Tolstoy function iff $-V(\mathbf{x})$ is a Tolstoy function. A simple example is $V(x) = (x - x^*)^2$ on $\Re$ with antimode x^*. Theorem 6.4 has an immediate corollary for reverse Tolstoy functions. Noting that here $S(u) = \{x : V(x) \leq u\}$, the proof is straightforward and can be omitted.

Corollary 6.1 *Let $A(u)$ be twice differentiable Then a sufficient condition that the mode of the density $f(x)$ coincides with the antimode of $V(x)$ is that $\phi(u)$ be strictly monotone decreasing, which holds if*

$$\frac{g'(u)}{g(u)} - \frac{A''(w)}{A'(u)} < 0, \quad u > 0 \ .$$

Example 6.5 The standard normal density can now be analyzed in terms of the reverse Tolstoy function $V(x) = x^2$ with antimode of zero. Here, $V(x) = u$ has the two solutions $\pm u^{\frac{1}{2}}$, so that $A(u) = 2u^{\frac{1}{2}}$ and $A'(u) = u^{-\frac{1}{2}}$. We also have $f(x) = \frac{1}{\sqrt{2\pi}} e^{-\frac{V(x)}{2}} = \phi(V(x))$, where $\phi(u) = \frac{1}{\sqrt{2\pi}} e^{-\frac{u}{2}}$. It follows that $g(u) = \frac{1}{\sqrt{2\pi}} u^{-\frac{1}{2}} e^{-\frac{u}{2}}$, which can be recognized as the familiar Chi-square result. Here, it can be seen directly that $\phi(u)$ is a monotone decreasing function.

6.1.2   *Weak Consensus and Accuracy*

As an application of the above results, we consider the question of the sufficiency of what may be called *weak consensus* for accuracy. The question of necessity of consensus for accuracy has been studied by Einhorn (1974), who showed that consensus of experts was a necessary condition for accuracy in a medical diagnosis context.

Let $V(\mathbf{x})$ be a performance measure associated with a decision vector $\mathbf{x}$. For example, an academic department may wish to decide the best priorities for research, x_1, service, x_2, and teaching x_3 for use in its policy making, as was discuss in Chapter 1. Suppose it is accepted that there exists a $V(\mathbf{x})$, which if known, would provide a good model for the optimal decision, $\mathbf{x}^*$. However, due to complexity of the situation, time required, or cost considerations, it is deemed impractical to attempt a direct modeling of $V(\mathbf{x})$. In this case, the individuals of the group may provide their respective best estimates $\mathbf{x}^t$ of $\mathbf{x}^*$. This raises the following two questions. In what way should the $\mathbf{x}^t$-estimates be aggregated? And under what condition will such an aggregate coincide with $\mathbf{x}^*$.

To make these ideas more precise, consider

Definition 6.2:
Let $\mathbf{x}^t, t = 1, \cdots, T$, be expert estimates, which are independent and identically distributed according to the density $f(\mathbf{x})$. Let $\mathbf{x}^0$ be the mode of

$f(\mathbf{x})$. Then the experts are said to have a weak consensus on $\mathbf{x}^0$. The question to be considered is under what conditions does $\mathbf{x}^* = \mathbf{x}^0$. That is, when will the weak consensus of experts be accurate?

To apply the previous results, let $g(v)$ be the performance density from which the individual $V^t = V(x^t)$ may be considered as independent, identically distributed observations. Also, we assume $V(\mathbf{x})$ is a Tolstoy function for which $A'(u) < 0$ for $0 \le u \le v_{\max}$. Then Theorem 6.4 provides a sufficient condition that the mode aggregate of the $\mathbf{x}^t$ identifies $\mathbf{x}^*$. Similarly, if the problem is considered in terms of a reverse Tolstoy function then Corollary 6.1 results apply. Namely, a sufficient condition of this type holds if $f(\mathbf{x})$ is an appropriate monotone function of $V(\mathbf{x})$.

Two approaches to using these results can be proposed. First, the group of experts or an independent analyst can propose a suitable $V(\mathbf{x})$-function, such as the reverse Tolstoy function, $V(\mathbf{x}) = k\|\mathbf{x} - \mathbf{x}^*\|^2$, which is the classical squared error loss function. Then information on $g(v)$ might be deduced from the sample $\{\mathbf{x}^t\}$, and tested in light of Corollary 6.1. Second, a new estimation technique based on maximizing the average v-score is proposed in Chapter 7.

In conclusion, a class of functions of interest for optimization and consensus was defined and characterized. Study of this class is motivated by its relation to a quote of Tolstoy and similar popular observations. A concept of weak consensus of experts was developed. Sufficient conditions were derived for the weak consensus of experts to identify an accurate or most desirable decision as the mode-aggregate of a set of expert estimates of the ideal decision.

6.2   Normal-Like Performance on Finite Intervals

This section discusses the modeling of random quantities with values in a finite interval, without loss of generality, the interval $[0, 1]$. In input-output efficiency analysis, efficiency scores are constrained to this interval. Data envelopment analysis (DEA) comprises a popular set of such input-output efficiency models. See Charnes *et al.* (1994), Coelli *et al.* (1998), Troutt (1997), Troutt *et al.* (2000), (2001) and (2002). Closely related are the frontier and stochastic frontier estimation or regression models. See Aigner and Chu (1997), Coelli *et al.* (1998), Troutt *et al.* (2000) and Troutt *et al.* (2002). In these kinds of models, efficiency shortfalls are typically nonnegative values in the range $[0, \infty)$ and can be transformed to the unit interval with the negative exponential function. Also, test scores in the range of 0-100% can be viewed as unit interval scores.

The only other approach known to the authors for construction of a normal-like density on a finite interval is the maximum entropy solution subject to a given mean and variance. In this chapter, we propose a new approach based on VDR. The resulting pdf model class is derived and compared with the maximum entropy approach. We argue that the maximum entropy approach is less appropriate in the presence of the purposeful behavior being modeled. Also, the new model class permits the modeling of distributions with unbounded concentration at the mode.

6.2.1 *Entropy-Based Generalizations of the Normal Density*

The maximal entropy characterization of the normal density and related results are not widely available. The only discussion known to us is in Reza (1961). The entropy of a pdf f(x) defined on [a,b] is given by

$$\int_a^b -f(x)\ln f(x)dx$$

It is shown there that:

(i) For the unit interval, the pdf that maximizes entropy is the uniform.
(ii) For $f(x)$ defined on $[0, \infty)$ with a specified mean $\mu > 0$, the maximizing pdf is the negative exponential with mean μ.
(iii) For $f(x)$ defined on $(-\infty, +\infty)$ with mean 0 and a specified variance, the pdf that maximizes entropy is the normal density with mean zero and the specified variance.

To illustrate the derivation of these results and obtain some additional results, let us consider the following problem. Find the pdf defined on $[a, b]$ having mean μ and variance σ^2 which maximizes entropy. Stated precisely, the problem is

$$\max \int_a^b -f(x)\ln f(x)dx$$

$$\text{subject to} \quad \int_a^b f(x)dx = 1$$

$$\int_a^b xf(x)dx \;=\; \mu$$

$$\int_a^b (x-\mu)^2 f(x)dx \;=\; \sigma^2$$

$$f(x) \;\geq 0.$$

Variational problems of this kind can be solved by formulating a Lagrangian functional, and applying the Euler-Lagrange Equation (Weinstock, 1952). Let λ_0, λ_1 and λ_2 be Lagrange multipliers for the three integral constraints, respectively. Define the Lagrangian L, by

$$L = -f(x)\ln f(x) - \lambda_0 f(x) - \lambda_1 x f(x) - \lambda_2 (x-\mu)^2 f(x). \qquad (6.14)$$

The Euler-Lagrange necessary condition for the function, $f(x)$, to provide an extremal solution in this setting is that $\frac{\partial L}{\partial f} = 0$. This yields the requirement that

$$-1 - \ln f(x) - \lambda_0 - \lambda_1 x - \lambda_2 (x-\mu)^2 = 0. \qquad (6.15)$$

Solving for $f(x)$ and collecting constants, the solution is thus of the form

$$f(x) = C \exp\{\lambda_1 x + \lambda_2 (x-\mu)^2\}. \qquad (6.16)$$

where C is the density constant. This class can be seen to be composed of products of the truncated exponential and truncated normal density function forms. Special cases of interest are first that in which the variance is not specified ($\lambda_2 = 0$) and the interval is $[0,1]$. The solution in that case is the positive or negative exponential depending on whether the mean is less than or greater than 0.5. A second special case is that for which $\lambda_1 = 0$. In that case, the solution is a truncated normal density function and λ_2 may be regarded as given by $\lambda_2 = \frac{1}{(2\sigma^2)}$. Shapes of the truncated normal pdf are illustrated in Johnson and Kotz (1987) for the case of $\mu = 0$. These shapes are similar to those of the new pdf class proposed in this section. However, numerical integration is generally required for obtaining the density constants, which limits their practical usefulness as compared to the newly proposed class.

6.2.2 *Normal VDR*

Consider the normal density on $\Re$ given by $f(x) = (2\pi\sigma^2)^{-1/2}\exp\{-(x-\mu)^2/2\sigma^2\}$. Let us regard $V(x) = (x-\mu)^2$ as the performance measure for efforts aimed at achieving $x = \mu$ on each trial. Theorem 1.2 (see Remark) may be used to obtain the corresponding $g(v)$ density as follows. We have

$$A(v) = L\{x : (x-\mu)^2 \le v\} = 2v^{1/2}. \tag{6.17}$$

Hence, $A'(v) = v^{-1/2}$ and

$$\begin{aligned}
g(v) &= \phi(v)A'(v) \\
&= f(x(v))A'(v) \\
&= (2\pi\sigma^2)^{-1/2}v^{-1/2}\exp\{-v/2\sigma^2\} \quad .
\end{aligned} \tag{6.18}$$

This density will be recognized as the gamma density, $\mathcal{G}(\alpha,\beta)$, with $\alpha = \frac{1}{2}$ and $\beta = 2\sigma^2$. Of course, this result is well known but we wish to use the present method as a technique for deriving the desired density class on $[0,1]$. In particular, this characterization shows that normal observations on $\Re$ may be regarded as arising by the following process. A target value of x, say x^*, is given. The test subject attempts to provide a performance as close to x^* as possible. If this performance is measured by $V(x) = (x-x^*)^2$, and the two (in general) x-values associated with a performance value of v are equally likely, then the performance measure, v, is distributed as $\mathcal{G}(\frac{1}{2},\beta)$ for some $\beta = 2\sigma^2$.

6.2.3 *Normal-like Performance on* $[0,1]$

The foregoing results suggest that if squared error is the performance measure, then the normal density of results occurs when the performance density, $g(v)$, is a gamma density with $\alpha = \frac{1}{2}$. Assuming that $x^* = 1$ is the target value on the interval $[0,1]$, then by Theorem 6.1, a squared error type of performance measure, along with gamma performance will endow the resulting $f(x)$ with a normal-like property. Thus, we say that a performance measure with range $[0,\infty)$ is normal-like if its distribution has a gamma density.

Let $x \in [0,1]$ be the result for a test subject who is attempting to achieve $x^* = 1$. Clearly $(x-1)^2$ is a squared error measure, but is bounded due to the zero endpoint of the interval. However, the measure, $-\ln x$ has value zero at $x = 1$ and range $[0,\infty)$. Also, $-\ln x$ is approximately equal to

$1 - x$ in the vicinity of $x = 1$ by Taylor's series. This suggests that a good approximate squared error type measure is

$$V(x) = (-\ln x)^2, \tag{6.19}$$

since this $V(x)$ approximates $(1 - x)^2$ in a neighborhood of the target value $x = 1$, and has range $[0, \infty)$.

We note that for this performance measure, there is only one value of x, corresponding to a given value of $V(x)$. Namely,

$$x(v) = \exp\{-v^{1/2}\} \tag{6.20}$$

and therefore, the uniform conditional density, $h(x|v)$, is a point mass at $x(v)$. Furthermore, by inspection of the graph and noting that we are only concerned with $\ln x$ values for $x \leq 1$,

$$A(v) = 1 - \exp\{-v^{1/2}\}, \tag{6.21}$$

and therefore

$$A'(v) = 1/2v^{-1/2}\exp\{-v^{1/2}\} \quad . \tag{6.22}$$

Use of Theorem 1.2 with $g(v)$ as the $\mathcal{G}(\alpha, \beta)$ density yields

$$f(x) = 2(\Gamma(\alpha)\beta^\alpha)^{-1}(-\ln x)^{2(\alpha-1)}\exp\{-\ln x - (-\ln x)^2/\beta\} \tag{6.23}$$

$$= 2(\Gamma(\alpha)\beta^\alpha)^{-1}x^{-1}(-\ln x)^{2(\alpha-1)}\exp\{-(\ln x)^2/\beta\}. \tag{6.24}$$

The following observations about this density, which we call the *NLOB* density, may be checked.

(a) For all $\alpha > 0$, $\beta > 0$, $\lim_{x \to 0} f(x) = 0$.

(b) If $\alpha = \frac{1}{2}$, $f(1) = 2(\Gamma(\alpha)\beta^\alpha)^{-1} = 2(\pi\beta)^{-1/2}$.

(c) If $\alpha < \frac{1}{2}$, $\lim_{x \to 1} f(x) = +\infty$. This shows that $f(x)$ is distinct from the truncated normal class, since that class could only have $f(1)$ between zero and $(2\pi\sigma^2)^{-1/2} = (\pi\beta)^{-1/2}$; that is, the normal pdf could only be bounded.

(d) If $\alpha > \frac{1}{2}$, $\lim_{x \to 1} f(x) = 0$. Thus $f(x)$ has its mode in the interior of $[0, 1]$ if $\alpha > \frac{1}{2}$.

In Figure 6.1, we show the NLOB density for $\beta = 1$ and $\alpha = 5.0, 2.0, 1.0, 0.5$ and 0.25, in left to right order.

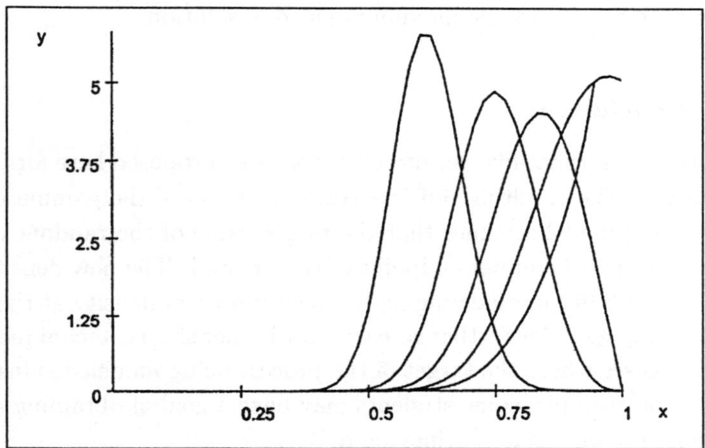

Figure 6.1: The NLOB density for $\beta = 1$ and
$\alpha = 5.0, 2.0, 1.0, 0.5$ and 0.25, in left to right order

The parameter α might be called the *modal intensity* parameter. As α increases from near zero to $\frac{1}{2}$, the mode of $f(x)$ remains at 1.0, where $f(x)$ is unbounded. When α becomes larger than $\frac{1}{2}$, the mode moves toward positive values. We regard this as a useful descriptive property of actual test scores. Consider a data set of examination scores with multiple 100% ($x = 1.0$) values. If these scores are slightly less than 1.0, but too close for discrimination from 1.0 or each other, then they may be considered as reflective of the high modal intensity case at 1.0. Hence, such a data set might well be modeled using $\alpha < \frac{1}{2}$. This is a clear advantage over the truncated normal class for a frequently occuring characterization of real test data sets.

Parameter estimation may be carried out according to existing methods for the $\mathcal{G}(\alpha, \beta)$ density after transforming the x_i data to $(\ln x_i)^2$. See, for example, Law and Kelton (1982) and also Chapter 7 below.

It may be argued that a maximum entropy solution is not appropriate in the test score context. Interpreting entropy as a measure of diversity or uncertainty, suggests that the maximum entropy solution would be most appropriate if the test subjects seek to maximize their diversity subject to a given mean and variance. However, we assume that the goal of each performance trial is to be as near to 1.0 as possible. Since the derived class is not the maximum entropy solution, it must have less entropy for its mean and variance than does the maximum entropy solution.

6.2.4 *Conclusion*

An alternative normal-like density class has been proposed here for the unit real interval. The resulting pdf is a transformation of the gamma density. The proposed model assumes that the target value of the random variable being modeled is the upper endpoint of the interval. The new density class permits a flexibility in modeling high concentration of density at the mode. An interesting issue for further research is whether similar techniques could be used in cases where the target of the process being modeled is inside the interval. For example, some students may have a goal of obtaining at least a passing score or a goal a achieving a "C".

6.3 Unimodality on the Half-Interval

Here we apply VDR techniques to monotone decreasing pdfs on the half-interval. Associated with a given such pdf, there are two others, the vertical density and the other one that we call the *dual* density. The inverse function of such a pdf is itself a pdf and may be considered as a *primal* pdf with its own vertical and dual densities. A theorem connecting these six pdfs is obtained. Next, Khintchine's Unimodality Theorem is revisited in light of these results. We also propose what we call Khintchine Density Representation (KDR). Finally, we give some remarks and an example on what we have called strata shift densities in Chapters 2 and 5. The strata shift operation permits any continuous pdf to be associated with a unique monotone decreasing pdf on the half-interval.

If $y = f(x)$ is a monotone decreasing pdf on the interval $[0, \infty)$, then so is $x = f^{-1}(y)$ (see, for example, Devroye, 1987). Using VDR, another two pdfs may be associated to both these pdfs. We consider their relationships and their application to unimodality. Call $f(x)$ the *primal* pdf. By Theorem 1.1, we are able to find the pdf of $y = f(x)$ and $g(v)$, the vertical density. An additional pdf, that we call the *dual* pdf, also arises in VDR. The first question is whether $g(y)$ and $f^{-1}(y)$ are in fact the same pdf. We show by an exponential pdf example that these densities do not generally coincide. Next, we find both vertical and dual pdfs for $f^{-1}(y)$ and relate these to the previous questions.

Recall Theorem 1.1 that the density for $V(x) = f(x)$ is given by $g(v) = -vA'(v)$. Let $g(v)$ be called the vertical pdf for $f(x)$. It was noted in Chapter 1 and can also be verified by integration of $g(v) = -vA'(v)$ that $A(v)$ is also a pdf. The pdf $A(v)$ will be called the dual density for $f(x)$.

6.3.1 *Relationships on the Half-Interval*

Let $f(x)$ be monotone strictly decreasing on $[0, \infty)$. Given a primal pdf $f(x)$ we may then consider forming the dual, inverse and vertical pdfs. The next theorem shows some relationships among these densities.

Theorem 6.5 *If $f(x)$ and $A(v)$ are differentiable monotone strictly decreasing functions, then we have*

(i) $A(v) = f^{-1}(v)$ *and* $A^{-1}(x) = f(x)$. *That is, the inverse and dual densities coincide.*

(ii) *The following relationships hold:*

Primal pdf	Dual pdf	Inverse pdf	Vertical pdf
$v = f(x)$	$A(v)$	$f^{-1}(v)$	$g(v) = -vA'(v)$
$x = f^{-1}(v) = A(v)$	$f(x)$	$f(x) = A^{-1}(x)$	$g(x) = -xf'(x)$

Proof.

(i) Since $v = f(x)$ is monotone decreasing, we note that the set $\{x : f(x) \geq v\}$ is the interval $[0, x]$ with Lebesgue measure x. Therefore,

$$A(v) = L\{x : f(x) \geq v\} = x = f^{-1}(v).$$

This can also be seen from the following considerations. Let $G(v)$ be the cumulative distribution function (cdf) for $v = f(x)$ and similarly, let $F(x)$ be the cdf corresponding to $f(x)$. Then $G(v) = 1 - F(f^{-1}(v))$. Therefore, $g(v) = -f(f^{-1}(v))\frac{d}{dv}f^{-1}(v) = -v\frac{d}{dv}f^{-1}(v)$.

Comparing with Theorem 1.1, we see that $A(v)$ and $f^{-1}(v)$ can differ at most by a constant. However, since both are densities that constant must be zero.

(ii) These follow easily from (i) and Theorem 1.1. We note that by considering inverses on each side of $A(v) = f^{-1}(v)$, we have $f(x) = A^{-1}(x)$.

Example 6.6   Let $f(x) = \exp(-x), x \in [0, \infty)$. Then, $A(v) = f^{-1}(v) = -\ln(v)$ and $g(v) = 1$ for $v \in [0, 1]$. One might expect that the vertical and inverse densities for $g(v)$ may also be among the previous densities. However, neither the vertical density nor the inverse density for $g(v)$ exist. This example also shows that the vertical and inverse pdfs are not identical.

In the next section, we consider these concepts in light of Khintchine's Unimodality Theorem.

6.4   Unimodality

Theorem 6.6   *A necessary and sufficient condition that a differentiable function, $f(x)$, be a monotone decreasing pdf function on $[0, \infty)$ is that $f(x)$ be the dual pdf for some pdf on $(-\infty, \infty)$.*

Proof: For sufficiency, suppose that for some pdf, $h(w)$, on $(-\infty, \infty)$ that $f(x)$ is its dual pdf, i.e.

$$f(x) = L\{w : h(w) \geq x\}.$$

Then, $f(x)$ is clearly a monotone decreasing pdf on $[0, \infty)$. For necessity, let $f(x)$ be a monotone decreasing pdf function on $[0, \infty)$. Then by Theorem 6.6-(ii), $f(x)$ is the dual of its inverse pdf.

The next theorem is a classical result due to Khintchine (1938); see also Feller (1971), Kotz and Johnson (1989), Devroye (1986) and Dharmadhikari and Joag-dev (1988).

Theorem 6.7   *(Khintchine's Unimodality Theorem): A necessary and sufficient condition for the distribution of a continuous random variable, X, to be unimodal about zero is that $X = UZ$, where U and Z are mutually independent random variables, and where U has the uniform pdf,*

$U[0,1]$. *If the probability density functions $f_z(z)$ and $f_x(z)$ exist with $f_x(z)$ differentiable, then*

$$f_z(z) = -z f_x'(z). \qquad (6.25)$$

Remark 1: Letting $f(x)$ denote the density of X and $g(x)$ denote the density of Z then we have $g(x) = -xf'(x)$. Thus, since $f(x)$ is the dual density for $f^{-1}(x)$, the density of the Khintchine random variable, Z, is the vertical density of $f^{-1}(x)$.

A proof of the Khintchine formula, $g(x) = -xf'(x)$, can be given with the help of Fig. 6.2, and leads to what may be called *Khintchine density representation* (KDR).

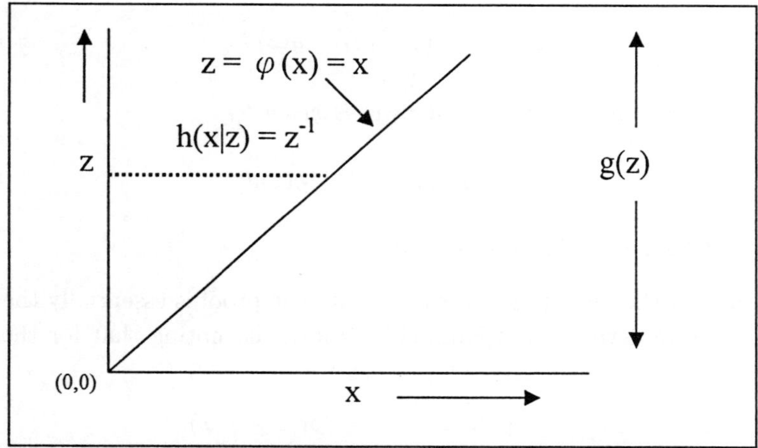

Figure 6.2: Graphical Depiction of Khintchine density representation

If $X = UZ$ with U uniform on $[0,1]$ as in Theorem 6.7, then given $Z = z$, the conditional density of X is uniform on $[0,z]$. Thus,

$$h(x|z) = \begin{cases} z^{-1} & \text{for } z \geq x, \\ 0 & \text{for } z < x \end{cases}$$

and

$$\begin{aligned} f(x) &= \int_0^\infty h(x|z)g(z)dz \\ &= \int_x^\infty z^{-1}g(z)dz. \end{aligned} \qquad (6.26)$$

Differentiating yield $f'(x) = -x^{-1}g(x)$, from which the Khintchine formula follows.

From this viewpoint, it is straightforward to generalize that result to cases in which $z = \varphi(x)$ is a more general monotone increasing or decreasing function. Let $z = \varphi(x)$ be a monotone function on $[0, \infty)$. Let $x = \varphi^{-1}(z)$ be the inverse function. Consider the following process, which may be called *Khintchine density representation* (KDR). Let z be distributed according to the density $g(z)$, $z \in [0, \infty)$. Then given z, let x be distributed according to the uniform density on $[0, \varphi^{-1}(z)]$. We have the following result for the density of the resulting random variable:

Theorem 6.8 *If $\varphi(x)$ is increasing, then the density of x is given by*

$$f(x) = \int_{\varphi}^{\infty} (\varphi^{-1}(z))^{-1} g(z) dz. \tag{6.27}$$

If $\varphi(x)$ is decreasing, then the density is given by

$$f(x) = \int_{0}^{\varphi} (\varphi^{-1}(z))^{-1} g(z) dz, \tag{6.28}$$

provided the required integrals exist.

Proof. For the monotone increasing case, the proof is essentially the same as the above derivation of Khintchine's formula, noting that for these assumptions,

$$h(x|z) = \begin{cases} (\varphi^{-1}(z))^{-1} & \text{for } z \geq \varphi(x), \\ 0 & \text{for } z < \varphi(x). \end{cases}$$

For the monotone decreasing case, we note that

$$h(x|z) = \begin{cases} 0 & \text{for } z \geq \varphi(x), \\ (\varphi^{-1}(z))^{-1} & \text{for } z < \varphi(x). \end{cases}$$

Remark 2: If $V(x) = x$ then (6.27) simplifies to

$$f(x) = \int_{x}^{\infty} v^{-1} g(v) dv \tag{6.29}$$

It follows that $f'(x) = -x^{-1}g(x)$ or $g(x) = -xf'(x)$. Thus Theorem 6.7 can be regarded as KDR applied to the function $V(x) = x$ and Theorem 6.8 gives a generalization of that result.

Some univariate examples for this Section are given next. Details of verifications are left to the reader. In each case, the conditional distribution of x given v is uniform on $[0, x(v)]$, as in Theorems 6.7 and 6.8.

Example 6.7 Let $V(x) = x$ and $g(v) = 1$ for $v \in [0, 1]$. Then $f(x) = -\ln(x), x \in [0, 1]$. We note that either Theorem 6.7 and 6.8 may be applied.

Example 6.8 Let $V(x) = x^2$ and $g(v) = 1$ for $v \in [0, 1]$. Then $f(x) = 1,\ x \in [0, 1]$.

Example 6.9 Let $V(x) = x^{1/2}$ and $g(v) = 1$ for $v \in [0, 1]$. Then $f(x) = x^{-1/2} - 1,\ x \in [0, 1]$.

6.5 Strata Shift Densities

Strata shifts were applied earlier in Chapters 2 and 5. Here we discuss this concept further and define *strata shift densities*. Consider an arbitrary continuous unimodal pdf f(x) on $(-\infty, \infty)$. It is possible to associate a unique monotone decreasing pdf on $[0, \infty)$ to $f(x)$, by what we call a strata shift. For each v in the range of $f(x)$, there are two zeroes of $v = f(x)$, which we denote as $x^-(v)$ and $x^+(v)$ in increasing order of values. We may construct a function, $s(x)$, as follows. For each v in the range of $f(x)$, define $w(v) = x^+(v) - x^-(v)$. Then define $\sigma(x)$ as the inverse function of $w(v)$. It follows that

$$L\{x : s(x) \geq v\} = L\{x : f(x) \geq v\} = x^+(v) - x^-(v).$$

Thus we have

Definition 6.3: Let $f_1(x)$ and $f_2(x)$ be pdfs defined on $\Re$. We say that $f_2(x)$ is a strata shift of $f_1(x)$ if the dual density of $f_2(x)$ is identical to that of $f_1(x)$. That is, each pdf has the same $A(v)$-function for $v \in [0, \infty)$.

The term, strata shift, suggests that horizontal slices or layers of $\{(x, y) : y \leq f(x)\}$ have been shifted until their lower ends align at the vertical axis. These densities have the same dual and vertical densities. Moreover, $s(x)$ is monotone decreasing on $[0, \infty)$ by Theorem 6.6. It therefore has an inverse function with its own dual and vertical density functions.

As an example, consider the standard normal pdf on the whole line, $f(x) = (2\pi)^{-1/2} \exp\{-\frac{x^2}{2}\}$. Here $A(v)$ is the measure of the interval $[-(-2\ln v - \ln(2\pi))^{1/2}, +(-2\ln v - \ln(2\pi))^{1/2}]$; namely, $A(v) = 2(-2\ln v -$

$\ln(2\pi))^{1/2}$. The same $A(v)$ can be obtained on the half-interval by using intervals of the form

$$[0, \ 2(-2\ln v - \ln(2\pi))^{1/2}] = [0, \ x(v)].$$

Inversion of $x(v)$ yields the pdf $(2\pi)^{-1}\exp\{-\frac{x^2}{8}\}$ as the desired strata shift pdf. Evidently, the original normal pdf can be reconstructed, in the distribution sense, by simulation from the strata shift pdf, and then choosing the random endpoints of the associated centered strata.

We may note more generally, that if $f(x)$ is a continuous pdf on $(-\infty, \infty)$, then $A(v)$ is the measure of the union of intervals, so that a strata shift pdf might, in principle, be associated to any such pdf.

6.6　The Use of the General VDR Theorem

6.6.1　*An Inverse Linear Programming Problem*

In Troutt, Tadisina and Sohn (2002) and Troutt, Pang and Hou (1999), the problem of fitting a linear programming model to observed input-output data is considered. Output data vectors $\mathbf{y}^t$ and input data vectors $\mathbf{x}^t$ are given for each of $t = 1, \ldots, T$ periods. The assumption is that the observed $\mathbf{y}^t$-vectors are attempts at optimal solutions $\mathbf{y}^{t*}$ of the linear programming models P_t given by maximize $\mathbf{c}'\mathbf{y}$ subject to $\mathbf{A}\mathbf{y} \leq \mathbf{x}^t$, where the components of $\mathbf{y}$ are nonnegative. However, the parameters, vector $\mathbf{c}$ and matrix $\mathbf{A}$ are not known, or perhaps not completely known. Fig. 6.3 depicts the situation in $\Re^2$ for one particular period t. As part of a Maximum Likelihood Estimation (MLE) solution strategy, the pdf-value for the observation $\mathbf{y}^t$ is required. By a separate procedure, the efficiency score $v \in [0, 1]$ for $\mathbf{y}^t$ is estimated, where $\mathbf{c}'\mathbf{y}^t = vz^*$ and z^* is the optimal objective function value for the linear programming problem for that period. The pdf for the $v^t, g(v)$, is estimated from the data for all the periods using the method in Troutt (1995), which is also discussed in Chapter 8 below. It is necessary to obtain the density value $f(\mathbf{y}^t)$ under the assumption that the conditional pdf of $\mathbf{y}^t$ given v is uniform on the set $S(V(\mathbf{y}^t), \mathbf{x}^t)$. This set is the intersection of the feasible region $\mathbf{A}\mathbf{y} \leq \mathbf{x}^t$ with the hyperplane $\mathbf{c}'\mathbf{y}^t = vz^*$.

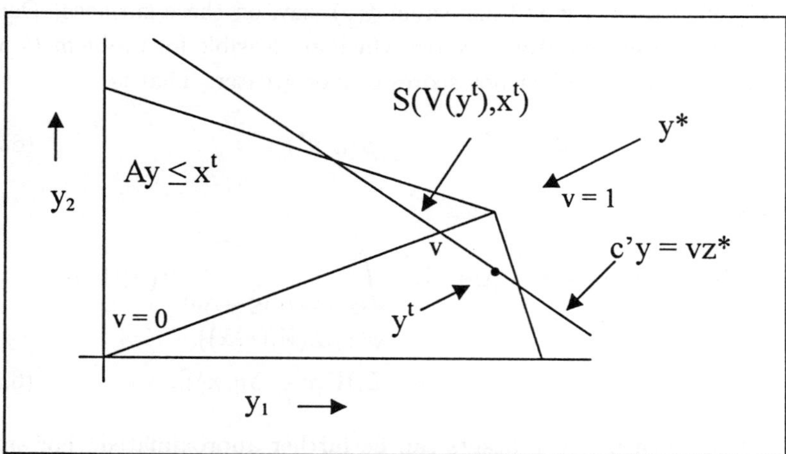

Figure 6.3: The Inverse Linear Programming Problem

From Theorem 1.5 (General VDR Theorem), we have $f(\mathbf{y}) = g(v) \parallel \nabla V(\mathbf{y}) \parallel h(\mathbf{y}|\ V = v)$. For this application, we have $V(\mathbf{y}) = (z^*)^{-1}\mathbf{c}'\mathbf{y}$. Hence, $\parallel \nabla V(y) \parallel = \parallel (\mathbf{z}^*)^{-1}\mathbf{c}' \parallel$. The conditional pdf $h(y|V = v)$ is the uniform pdf. Therefore, $h(\mathbf{y}|V = v) = (L\{S(V(\mathbf{y}^t),\mathbf{x}^t)\})^{-1}$. In Fig. 6.3, $L\{S(V(\mathbf{y}^t),\mathbf{x}^t)\}$ is the length of the line segment indicated. Thus, applying the theorem, we see immediately that

$$f(\mathbf{y}^t) = \parallel \mathbf{c} \parallel g(v)[z^* L\{S(V(\mathbf{y}^t),\mathbf{x}^t)\}]^{-1}.$$

6.6.2 *Comparison to a First Principles Approach*

The foregoing may be compared with a direct approach as was used in Troutt, Tadisina, Sohn and Brandyberry (2003c) and Troutt, Pang and Hou (1999). The pdf models $f(\mathbf{y}^t)$ for the feasible sets of problems P_t respectively;, represent the composition of a two-step process. First, a value v is selected according to the density $g(v)$. Then given v, a vector $\mathbf{y}$ is selected on the surface $S(V(\mathbf{y}^t),\mathbf{x}^t)$, according to the uniform density on that set. Let Δv be a small positive number and consider the approximation of the probability, $P(v \leq V(y) \leq v+\Delta v)$ in two ways. First, this probability is given by

$$\int_v^{v+\Delta v} g(u)du \doteq g(v)\Delta v.$$

By the uniform density assumption, $f(\mathbf{y})$ is constant on $S(V(\mathbf{y}^t), \mathbf{x}^t)$ for each v and has value $\phi(v)$, say, on these surfaces. Define $W(v, \mathbf{x}^t)$ as the set of output vectors which are feasible for problem P_t and which have decisional efficiency scores of v or greater. That is

$$W(v, \mathbf{x}^t) = \cup_{u \geq v} S(u, \mathbf{x}^t). \tag{6.30}$$

It follows that

$$\begin{aligned}
\Pr(v \leq V(\mathbf{y}) \leq v + \Delta v) &= \int_{\{\mathbf{y}: v \leq V(\mathbf{y}) \leq v + \Delta v\}} f(\mathbf{y}) \Pi dy_i \\
&\cong \phi(v)[L(W(v, \mathbf{x})) \\
&\quad - L(W(v + \Delta v, \mathbf{x}))].
\end{aligned} \tag{6.31}$$

The volume measure in brackets can be further approximated. For small Δv, it is given by the product of the surface measure $L(S(v, \mathbf{x}))$ and the distance element $k\Delta v$ corresponding to Δv and orthogonal to $S(v, \mathbf{x})$. Let $\mathbf{y}^*$ be the optimal solution for problem P_t with optimal objective function value z^*. The required distance is the length of the projection of the vector, $(\Delta v)\mathbf{y}^*$ in the direction of vector $\mathbf{c}$, which is

$$\frac{(\Delta v)z^*}{\| \mathbf{c} \|}. \tag{6.32}$$

It follows that

$$\begin{aligned}
&L(W(v, \mathbf{x})) - L(W(v + \Delta v, \mathbf{x}) \\
&\cong L(S(v, \mathbf{x})(\Delta v)z^* / \| \mathbf{c} \| .
\end{aligned} \tag{6.33}$$

Combining results, we have in the limit as $\Delta v \to 0$,

$$g(v) = \phi(v)L\{S(v, \mathbf{x})\}z^* / \| \mathbf{c} \| . \tag{6.34}$$

Therefore

$$f(y) = \varphi(V(\mathbf{y})), \tag{6.35}$$

where

$$\phi(v) = \| \mathbf{c} \| g(v)[z^* L(S(v, \mathbf{x}))]^{-1}. \tag{6.36}$$

To summarize, the density value for output vector $\mathbf{y}^t$ in problem P_t is given by

$$f(\mathbf{y}^t) = \| \mathbf{c} \| \, g(v)[z^* L(S(V(\mathbf{y}^t), \mathbf{x}^t))]^{-1}, \qquad (6.37)$$

as was obtained above more directly by VDR.

6.7 Conclusion

This chapter has illustrated four other applications of VDR. These techniques are useful for issues like unimodality. A concept called Tolstoy's Law was used to consider the relationship between unimodality of the score function $V(x)$ and unimodality of the density of estimates as related to the vertical density. Khintchine's Unimodality Theorem was discussed and led to what we called KDR. The general VDR Theorem 1.5 was also shown to provide an efficient density-modeling tool for a MLE estimation setting. In the next chapter, we consider a case application to a cost estimation problem in production-planning.

Chapter 7

Management Science Applications of VDR–II

Minimum Decisional Regret Estimation

One of the original motivations for VDR was to better understand the statistics of performance, or more precisely, performance scores. The $V(\mathbf{x})$-function may be considered a performance score and its distribution is then characterized by the density $g(v)$. In this chapter, we propose a parameter estimation principle that we call *minimum decisional regret* (MDR). This estimation approach is based on the $g(v)$-pdf when that density depends on one or more parameters. Here we focus on mathematical programming models depending linearly on costs or other parameters. The approach is illustrated with a production-planning model as an example. The method uses past actual planning data to estimate appropriate cost parameter values for a stipulated planning model. Already known costs can be accommodated within the estimation procedure using some simple adjustments to the procedure. Thus, the method permits estimation of costs that are otherwise difficult to estimate directly via accounting information or special studies.

Such techniques may be called *behavioral estimation* since they are based on the actual decision-making behavior or observed actions of managers or firms. We propose a new validation approach for this estimation principle, which we call the *target-mode agreement* criterion. This validation method uses VDR methods more directly and further illustrates its usefulness as a probability density function modeling tool.

The method is based on what we call *decisional regret*, which is a measure of the distance between actual plans and those that would have been

149

optimal for the costs being estimated. The approach will be illustrated for the context of production-planning and we next give some background on that problem.

7.1　The Aggregate Production Planning Problem

The aggregate production-planning (APP) problem has historically been one of most thoroughly studied Operations Research/Management Science (ORMS) topics. The theory is well established in that effective methods have been proposed, demonstrated, implemented and compared. This theoretical maturity of the APP problem contrasts with actual practice. A number of textbook authors observe that use of mathematical models is the exception rather than the rule. See, for example, Peterson and Silver (1979), Silver *et al.* (1998), Stevenson (1982), Hax and Candea (1984), Schroeder (1993) and Vollman, Berry and Whybark (1997). In a survey of 500 companies in the Southeast U.S., Dubois and Oliff (1991) observed that few of the firms used any models at all, citing as a contributing reason, the considerable effort to obtain the required cost information. For other views on the lack of practitioner use of these models, see Nam and Logendran (1992) and Buxey (1995).

The Dubois and Oliff (1991) survey on practice reported that 64% of respondents did use a formal approach. However, few reported using any of the available cost-based mathematical programming models. It may be concluded that only the simplest models, such as graphical and spreadsheet analyses, were the most common formal methods. Spreadsheet modeling is further emphasized by Vollmann, Berry and Whybark (1992). The balance of respondents reported using extrapolation of previous decisions, or rules of thumb, with 5% not responding. We conjecture that a more up-to-date survey would indicate increased use of models today due to the convenient solvers that are available in spreadsheet software.

Another interesting result of the Dubois and Oliff (1991) study was that respondents generally rated the standard APP cost categories as very important for planning. However, very few actually estimated these costs. For example, shortage costs were rated second in importance; while only 9% reported estimating such costs. Their study concluded that the lack of cost estimates was an important impediment to the use of mathematical programming models and called for development work in this area that has so far not been met. Cost estimation for the APP problem not only requires effort, but is also difficult for at least the following reasons:

(i) The appropriate allocation of indirect costs may not be obvious.

(ii) The definition of cost is not always well specified. Examples of various definitions are: (a) total cost, (b) average cost, and (c) incremental cost.

(iii) Total costs are tied to a specific model and decision purpose.

(iv) Some APP costs may be impractical to estimate directly, e.g. back-orders and lost sales.

(v) Some costs may be regarded as implied by policy choices or actual behavior as in the exchange curve approach to aggregate inventory control (Brown 1961, 1967, Peterson and Silver, 1979). See also the discussion on optimal policy curves in Starr and Miller (1962).

In addition, costs can be regarded as stochastic (Lau and Lau, 1981, 1983, and Shih, 1981). However, in this chapter, costs are regarded as deterministic.

Thus, use of mathematical programming models for production planning in practice has not been enthusiastic with evidence that the lack of suitable cost estimation systems is an important limitation. Information on costs comes from several sources such as accounting data, opinions and managerial policy. However, information on costs can also be obtained from the firm's past decisions. The method proposed here provides for an integration of these sources.

Additional motivation for this work can be also conveyed with the help of a footnote remark by Peterson and Silver (1979). Speaking of Bowman's (1963) *Management Coefficients Theory* (MCT), they write:

Strictly speaking one cannot really compare actual company decision behavior with a management coefficients model, or with any of the other explicit models discussed in this chapter. The objective function which management may have tried to minimize with their past decisions may have been far different from any of the ones we have formulated mathematically.

While the truth of this is self-evident, the following question can nevertheless be posed. Based on a measure of purposeful action, which model in a class of models renders the past decisions of the firm most plausible? Thus, it may be true that the firm had no explicit model in mind at all. However, it is still of theoretical and managerial interest to reveal the implicit model(s) and costs that best explain past decisions.

In the next section, we will develop the proposed decisional regret method in the case that no cost parameters are known. For that situation, normalized relative costs can be estimated. We then consider how known costs can be accommodated. If any of the costs are assumed to be known, as should be the typical case in practice, then all the remaining costs can be estimated.

7.2 Minimum Decisional Regret Estimation of Cost Parameters

It is helpful to first consider the case in which none of the cost parameters is known. In that case, normalized values of all cost parameters can be estimated. Usually, the values of several cost parameters will be known with sufficient accuracy from other information sources such as accounting reports or special studies. We then show how the procedure can be modified to accommodate those cost parameters that are considered to be known.

7.2.1 *Conceptual Model of the Planning Process*

From the literature review above, the typical APP planning approach can be outlined as follows. Available data will generally include some key costs such as regular time and overtime time production and sub-contracting costs. According to Dubois and Oliff (1991), typical forecasts were based on trend lines or trend lines with seasonal adjustments. Thus, we assume that forecasts are available at least through the planning horizon, along with beginning inventory and workforce levels. Starting with these data, a simple graphical or spreadsheet model will likely have been used to evaluate various trial plans. Following recommended practice, such plans would have been discussed and modified according to judgment and other information to achieve a final plan. Using a rolling horizon, only the first period part of final plans would have been implemented. For the estimation approach proposed here, it is therefore assumed that a set of historical final plans are available, along with the forecasts and other initial conditions on which they were based. These should most typically be found in finalized spreadsheet planning models.

7.2.2 *Some Notational Conventions and Definitions*

Suppose p is the number of decision variables and let $\mathbf{x}^{t,h} \in \Re^p$ be the decision vector for planning period-$(t+h)$, which is made at period-t. See the chapter Appendix for instances of the components of these vectors in the example planning model given there. For example, the first coordinate of $\mathbf{x}$ is H, the number hired. There are $p = 8$ decision variables for each period in the planning horizon. We assume that data are given for $t = 1,\ldots,T$ periods. The standard rolling horizon assumption is made, in which at each planning period-t, the total expected cost for the horizon periods $t+h, h = 1,\ldots,N$, is to be minimized subject to a feasible region, S^t. Thus, S^t is specified by equalities and inequalities relating inventory levels, workforce levels, and the other decision variables and parameters to forecasts.

For the model in the chapter Appendix, S^t is defined by constraints, A.2-A.17 along with the parameter values selected. We have $S^t \subset \Re^k$, where the dimension, k, is discussed below. In what follows, we refer to the underlying mathematical programming model as the *stipulated model*.

We may define normalized implicit costs, ξ_i, such that $\xi_i \geq 0$, and $\sum_i \xi_i = 1$. These are multiplied by the terms corresponding to the costs, $c_i, i = 1,\ldots,n$, respectively, in the objective function of the model in the chapter Appendix. Let $f_i(\mathbf{x}^{t,h})$, which we call a cost driver, be the term multiplied by c_i, or equivalently, ξ_i. For that model, we have $n = 10$ as the number of costs and cost drivers. These terms are decision variables or functions of decision and system variables. As examples, $f_2(\mathbf{x}^{t,h}) = H^2_{t,h}$ and $f_{10}(\mathbf{x}^{t,h}) = Z_{t,h} - Z_{t,h-1}Z_{t,h}$ in the model in the chapter Appendix. It follows that the total expected normalized implicit costs at period-t can be expressed as

$$E_t(\boldsymbol{\xi}) = E_t(\boldsymbol{\xi}; \mathbf{x}^{t,1},\ldots,\mathbf{x}^{t,N}) = \sum_h \sum_i \xi_i f_i(\mathbf{x}^{t,h}) \tag{7.1}$$

This notation is abbreviated, omitting dependence on the forecasts and initial conditions given. For a still more simplified notation, we may observe that plans consist of N vectors of dimension p, one for each period in the planning horizon. Define $\mathbf{X}_t \in \Re^k$ with $k = p \times N$ as the row vector obtained by concatenating the row vectors, $x^{t,h}$. Next, define $\Psi_i(\mathbf{X}_t) = \sum_h f_i(\mathbf{x}^{t,h})$.

Thus, (7.1) may then be written as

$$E_t(\boldsymbol{\xi}) = E_t(\boldsymbol{\xi}; \mathbf{X}_t) = \sum_{i=1}^{n} \xi_i \Psi_i(\mathbf{X}_t) \qquad (7.2)$$

and the APP problem for period-t can be described concisely as :

$$\text{APP}_t \qquad \min E_t(\boldsymbol{\xi}) = E_t(\boldsymbol{\xi}; \mathbf{X}_t) = \sum_{i=1}^{n} \xi_i \Psi_i(\mathbf{X}_t), \quad \text{s.t.} \quad \mathbf{X}_t \in S^t. \quad (7.3)$$

We note that $\boldsymbol{\xi}$ is a fixed parameter vector in (7.3).

7.3 Decisional Regret

Let us denote the data for past plans by $\mathbf{X}_t^o$, for $t = 1, \ldots, T$. Under the rolling horizon assumption, only the first period portion of any actual past plan would have been implemented. Nevertheless, the other periods in past plans should be expected to provide relevant information. This is because all horizon periods in the plan are decided simultaneously. That is, the first period parts of such plans may be said to be optimal ones (i. e. optimally satisfactory to the planners) only in connection with the optimal solutions for the other periods in the horizon. The $\mathbf{X}_t^o$ are assumed to have been feasible for problem, APP_t; that is, we assume that $\mathbf{X}_t^o \in S^t$. This assumption is discussed further in Section 7.5 below.

Next, we denote by $\mathbf{X}_t^*$ the optimal solutions or plans for problems APP_t. In case there exist alternative optimal solutions for (7.3), then $\mathbf{X}_t^*$ may be chosen as any one of those alternatives. We call the $\mathbf{X}_t^* = \mathbf{X}_t^*(\boldsymbol{\xi})$ the *model-optimal* plans. For each historical planning period-t, define the *decisional regret* (DR), $\delta_t(\boldsymbol{\xi})$, by

$$\delta_t(\boldsymbol{\xi}) = E_t(\boldsymbol{\xi}; \mathbf{X}_t^o) - E_t(\boldsymbol{\xi}; \mathbf{X}_t^*(\boldsymbol{\xi})) \qquad (7.4)$$

Under the above assumptions, $\delta_t(\boldsymbol{\xi})$ can only be nonnegative. These δ_t may be regarded as measures of distance between an actual final plan and one that would have been optimal for the normalized implicit cost parameter vector and stipulated APP model.

We may now state the Minimum Decisional Regret (MDR) estimation principle as follows. Past management decisions or plans are assumed to have had the goal of minimizing total costs according to the stipulated model. This implies that the $\delta_t(\boldsymbol{\xi})$ should have been minimized for each t.

Therefore, over all periods, management decision-making should have been aimed at minimizing their sum or, equivalently, their average. Hence, the MDR principle proposes as the estimate of ξ, that vector, which minimizes $\sum_{t=1}^{T} \delta_t(\xi)$.

The MDR idea may be considered to have evolved from several earlier related concepts. In his development of Management Coefficients Theory (MCT), one of Bowman's (1963) assumptions was that managers make reasonably good decisions on average. More precisely, his assumption was that the bias in their decisions is expected to be small, but their variance may be unacceptably large and costly. This can be viewed as implying that managerial and perhaps other expert decision-making is generally expected to be close to optimal in some sense. Therefore, it is reasonable to think in terms of how far from being optimal such decisions may have been, to construct related measures of distance and to use validation tests for the assumption of approximately optimal decisions. Minimizing the total or average distance from optimal plans is consistent with Bowman's assumption. Thus, this method seeks the costs for which the associated model-optimal plans are as near to the actual plans as possible, on average.

In another related connection, Ackoff (1962) discussed a parameter estimation method he called retrospective optimization. By this, he meant that given a normative model for a decision vector that depends on a parameter and the actual decision that was selected, then one might solve for the parameter value that would have rendered the observed decision optimal for the model. A similar idea was used earlier also in the *exchange curve theory* of Brown (1961,1967). Stevenson (1982) also provides an exercise using retrospective optimization in connection with estimating costs in the Single-Period Inventory problem.

Exchange curve theory showed that assuming the Economic Order Quantity (EOQ) model (see, for example, Stevenson, 1982, and Troutt, 1991))as the stipulated optimization model, one could use total annual number of orders and average inventory levels to solve for the ratio of ordering cost to holding cost that would have rendered the aggregate decisions optimal. Peterson and Silver (1979) call this *imputation of costs* and attribute it to Starr and Miller (1962). More recently, the maximum decisional efficiency (MDE) technique of Troutt (1995) and a variation called maximum performance efficiency (MPE) estimation (Troutt *et al.* 2000) have been proposed. These develop ratio measures of actual performance to model-optimal performance as functions of unknown parameters. This in turn makes it similar to efficient frontier estimation models such as those of Aigner and Chu (1968).

The MDR approach is similar to those approaches but is better suited to estimating parameters in mathematical programming models. Validation of such estimates is discussed in Section 7.4 using the target-mode agreement criterion. These considerations lead to formulation of the MDR estimation problem given by

MDR:

$$\text{Minimize:} \quad \sum_{t=1}^{T} \delta_t(\boldsymbol{\xi}),$$

$$\text{s.t.} \quad \sum_{i=1}^{n} \xi_i = 1. \tag{7.5}$$

Theorem 7.1

    (i) $\delta_t(\boldsymbol{\xi})$ is a convex function of $\boldsymbol{\xi} \in \Re^n$ for each $t = 1, \ldots, T$.

    (ii) For a feasible $\boldsymbol{\xi}^o$, the affine linear function, $L^t(\boldsymbol{\xi}; \boldsymbol{\xi}^0)$, given by

$$
\begin{aligned}
L^t(\boldsymbol{\xi}; \boldsymbol{\xi}^o) &= E_t(\boldsymbol{\xi}; \mathbf{X}_i^o) - E_t(\boldsymbol{\xi}; \mathbf{X}_t^*(\boldsymbol{\xi}^o)) \\
&= \sum_i \xi_i \Psi_i(\mathbf{X}_t^o) - \sum_i \xi_i \Psi_i(\mathbf{X}_t^*(\boldsymbol{\xi}^o)) \\
&= \sum_i \xi_i \Psi_i(\mathbf{X}_t^o) - \Psi_i(\mathbf{X}_t^*(\boldsymbol{\xi}^o)) \\
&= \sum_i (\xi_i - \xi_i^o)(\Psi_i(\mathbf{X}_t^o) - \Psi_i(\mathbf{X}_t^*(\boldsymbol{\xi}^o))) \\
&\quad + \delta_t(\boldsymbol{\xi}^o)
\end{aligned}
\tag{7.6}
$$

    is a supporting hyper-plane to the graph of $\delta_t(\boldsymbol{\xi})$ at $\boldsymbol{\xi}^o$ for $t = 1, \ldots, T$.

    (iii) $\sum_{t=1}^{T} \delta_t(\boldsymbol{\xi})$ is a convex function of $\boldsymbol{\xi}$ with supporting hyper-plane at $\boldsymbol{\xi}^0$ given by $\sum_t L^t(\xi; \boldsymbol{\xi}^o)$, $t = 1, \ldots, T$.

Proof. First, we show that $w(\boldsymbol{\xi}) = E_t(\boldsymbol{\xi}; \mathbf{X}_t^*(\boldsymbol{\xi})) = \min \sum_{i=1}^{n} \xi_i \Psi_i(\mathbf{X}_t)$ s.t. $\mathbf{X}_t \in S^t$, is a concave function of $\boldsymbol{\xi}$. Let $\boldsymbol{\xi}^1$ and $\boldsymbol{\xi}^2$ be implicit cost vectors and let nonnegative λ_1 and λ_2 be given with $\lambda_1 + \lambda_2 = 1$. We have

$$w(\lambda_1 \boldsymbol{\xi}^1 + \lambda_2 \boldsymbol{\xi}^2) = E_t(\lambda_1 \boldsymbol{\xi}^1 + \lambda_2 \boldsymbol{\xi}^2; \mathbf{X}_t^*(\lambda_1 \boldsymbol{\xi}^1 + \lambda_2 \boldsymbol{\xi}^2)$$

$$
= \min \sum_{i=1}^{n} (\lambda_1 \xi_i^1 + \lambda_2 \xi_i^2) \Psi_i(\mathbf{X}_t)), \quad \text{s.t. } \mathbf{X}_t \in S^t
$$

$$
= \sum_{i=1}^{n} (\lambda_1 \xi_i^1 + \lambda_2 \xi_i^2) \Psi_i(\mathbf{X}_t')
$$

$$
= \lambda_1 \sum_{i=1}^{n} \xi_i^1 \Psi_i(\mathbf{X}_t') + \lambda_2 \sum_{i=1}^{n} \xi_i^2 \Psi_i(\mathbf{X}_t')
$$

$$
\geq \lambda_1 w(\boldsymbol{\xi}^1) + \lambda_2 w(\boldsymbol{\xi}^2) \tag{7.7}
$$

where $\mathbf{X}_t' = \mathbf{X}_t^*(\lambda_1 \boldsymbol{\xi}^1 + \lambda_2 \boldsymbol{\xi}^2)$. Therefore, $w(\boldsymbol{\xi})$ is concave. Since $E_t(\boldsymbol{\xi}; \mathbf{X}_t^o)$ is linear, hence convex, in $\boldsymbol{\xi}$, and since $-w(\boldsymbol{\xi}) = -E_t(\boldsymbol{\xi}; \mathbf{X}_t^*(\boldsymbol{\xi}))$ is convex, it follows that $\delta_t(\boldsymbol{\xi}) = E_t(\boldsymbol{\xi}; \mathbf{X}_t^o) - E_t(\boldsymbol{\xi}; \mathbf{X}_t^*)$ is convex and part (i) is established.

For part (ii), let $\boldsymbol{\xi}^o$ be a particular feasible implicit cost vector. We claim that the linear function of $\boldsymbol{\xi}$ given by $\sum_{i=1}^{n} \xi_i \Psi_i(\mathbf{X}_t^*(\boldsymbol{\xi}^o))$ is tangent to the graph of $w(\boldsymbol{\xi})$ at $\boldsymbol{\xi} = \boldsymbol{\xi}^o$. To see this, we have

$$
w(\boldsymbol{\xi}) = \min \sum_{i=1}^{n} \xi_i \Psi_i(\mathbf{X}_t^*) \quad \text{s.t. } \mathbf{X}_t \in S^t \tag{7.8}
$$

and therefore

$$
w(\boldsymbol{\xi}) \leq \sum_{i=1}^{n} \xi_i \Psi_i(\mathbf{X}_t^*(\boldsymbol{\xi}^o)) \tag{7.9}
$$

Thus, the graph of $\sum_{i=1}^{n} \xi_i \Psi_i(\mathbf{X}_t^*(\boldsymbol{\xi}^o))$ lies above that of $w(\boldsymbol{\xi})$. However, by the definition of $\mathbf{X}_t^*(\boldsymbol{\xi}^o)$, it also holds that $w(\boldsymbol{\xi}^o) = \sum_{i=1}^{n} \xi_i^o \Psi_i(\mathbf{X}_t^*(\boldsymbol{\xi}^o))$. Thus the hyper-plane, $\sum_{i=1}^{n} \xi_i \Psi_i(\mathbf{X}_t^*(\boldsymbol{\xi}^o))$, is tangent to the graph of $w(\boldsymbol{\xi})$ at $\boldsymbol{\xi} = \boldsymbol{\xi}^o$ and is therefore a supporting hyper-plane. Since $E_t(\boldsymbol{\xi}; \mathbf{X}_t^o)$ is a linear function of $\boldsymbol{\xi}$, it is its own tangent at every value of $\boldsymbol{\xi}$. It follows that $L_t(\boldsymbol{\xi}; \boldsymbol{\xi}^o) = E_t(\boldsymbol{\xi}; \mathbf{X}_t^o) - E_t(\boldsymbol{\xi}; \mathbf{X}_t^*(\boldsymbol{\xi}^o))$ is tangent to the graph of $\delta_t(\boldsymbol{\xi})$ at $\boldsymbol{\xi} = \boldsymbol{\xi}^o$ for each t. This establishes (ii). Since $L_t(\boldsymbol{\xi}; \boldsymbol{\xi}^o)$ is tangent to the graph of $\delta_t(\boldsymbol{\xi})$ at $\boldsymbol{\xi} = \boldsymbol{\xi}^o$ for each t, part (iii) clearly follows, thus, concluding the proof.

Problem MDR may be solved by the supporting hyper-plane method. See Veinott (1967), Avriel (1976), Bazaraa *et al.* (1993) or Bazaraa *et al.* (1990). That algorithm can be adapted as follows.

MDR Algorithm: Let Λ be a constraint list, which initially contains

Bazaraa, Jarvis and Sherali (1990). That algorithm can be adapted as follows.

MDR Algorithm: Let Λ be a constraint list, which initially contains only (7.5). Let ξ^o be an arbitrary feasible starting iterate such as $\xi_i^o = 1/n$ for $i = 1,\ldots,n$. Let ω be an unrestricted auxiliary variable. Let τ be a zero tolerance.

Step 0: Set $k = 0$ and let ξ^o be an arbitrary starting iterate feasible for (7.5).

Step 1: Solve problems APP_t, using $\xi = \xi^k$ and determine L^t for each $t = 1,\ldots,T$.

Step 2: Add to Λ the constraint $\sum_{t=1}^T L^t(\xi;\xi^k) \leq \omega$, or equivalently, $\sum_{i=1}^n \xi_j(\sum_{t=1}^T \{\Psi_i(X_t^o) - \Psi_i(X_t^*(\xi^k))\}) - \omega \leq 0$

Step 3: Solve the linear programming problem given by: *minimize* ω subject to the constraints in Λ. Let ξ^{k+1} be any solution.

Step 4: If $\| \xi^k - \xi^{k+1} \| < \tau$ then stop. Otherwise, advance k and iterate from Step 1.

This MDR approach may be compared to a more direct strategy, which would generally be intractable. If the model-optimal plans could be expressed as explicit functions of the normalized implicit cost vector as in retrospective optimization, then a regression model might be possible. However, in general, it is not possible to obtain explicit functions for the optimal solutions of mathematical programming problems. Figure 7.1 illustrates the approach of the algorithm.

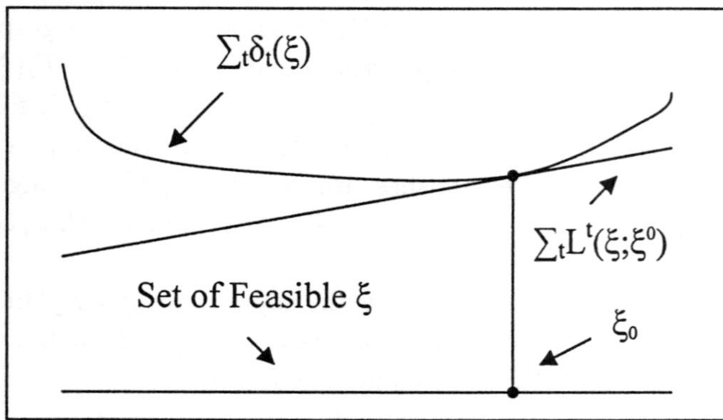

Figure 7.1:MDR Algorithm Strategy

7.3.1 *Handling of Costs Assumed to be Known*

In the foregoing, none of the cost parameters was assumed to be known. In that case, the MDR model could then estimate normalized, sum-to-unity, costs. Suppose that true costs are given by c_i with $\sigma = \sum_{i=1}^{n} c_i$ and $\xi_1^* > 0$. Then we assume $\xi_i^* = \frac{c_i}{\sigma}$. This solution is consistent with the assumption that if any particular cost is known then all of them may be obtained by proportionality. That is, if only $c_1 > 0$ is assumed to be known and $\xi_1^* > 0$, then we have $c_i = c_1(\frac{\xi_i^*}{\xi_1^*})$, for all $i = 1, \ldots, n$. The case of $\xi_1^* = 0$ is discussed below.

Next, suppose that two or more of the c_i are known. Let $c_i > 0$, $i \in I_o \subset \{1, 2, \ldots, n\}$, say, be the costs assumed to be known. We may then include in the MDR model constraints of the form $\frac{\xi_i}{\xi_j} = \frac{c_i}{c_j}$ for $i, j \in I_o$. Such constraints are easily converted to linear form. If the resulting estimates, $\xi_i^* > 0$ for $i \in I_o$, then using any one of them, all may be obtained by proportionality as above. We may note that ranges of such ratios, if known, might also be included as constraints.

In the case that $\xi_i^* = 0$ for any $i \in I_o$, for a cost known to be positive then we must obtain exogenous lower and upper bound estimates of the form, $0 < a_i < c_i < b_i$, for all the costs. It follows that $\sum_{i=1}^{n} a_i < \sigma < \sum_{i=1}^{n} b_i$, and therefore $a_i(\sum_{i=1}^{n} b_i)^{-1} \leq \xi_i^* \leq b_i(\sum_{i=1,n} a_i)^{-1}$. We note that $a_i = c_i = b_i$, for $i \in I_o$. Thus, the constraints, $a_i(b_i)^{-1} \leq \xi_i \leq b_i(\sum_{i=1}^{n} a_i)^{-1}$, for all $i = 1, \ldots, n$, would be added to those of the MDR estimation problem, ensuring that all estimated costs will be positive.

Thus, if any of the cost parameters are assumed to be known, which is the expected case in practice, then all of them can be estimated. The decisional regret values can then be expressed in dollars by substitution of the optimal cost estimates into (7.4). In the following, we use the notation, δ_t, for the estimated decisional regret values and assume that these are expressed in dollars when one or more costs are known. It should also be noted that $\sum_{t=1}^{T} \delta_t$, expressed in dollars or not, may be regarded as a measure of goodness-of-fit that might be used to compare competing models. See Figure 7.2.

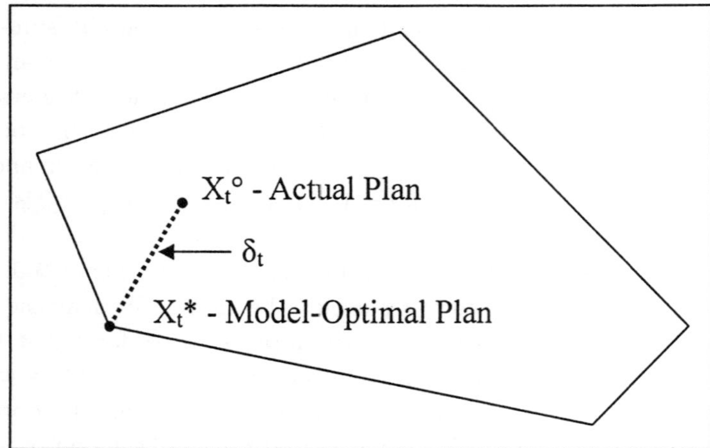

Figure 7.2: Period-*t* Feasible Set

7.4　Validation

Two kinds of validation issues arise here. First, for the stipulated APP model, we wish to assess whether there is evidence that management plans are consistent with the goal of minimization of costs according to that model. If not, then the basic rationale of the MDR estimation technique may not be met and the results may not be plausible. This parallels model aptness testing for regression models in MCT. See Madansky (1988). That is, the estimation technique might mechanically produce parameter estimates for almost any set of data, but possibly without evidence of validity of the assumptions. The second issue is what may be called routine model validation, in which model predicted costs can be compared with accounting or other information.

7.4.1　*The Target-Mode Agreement Criterion*

In this section, we propose a validation criterion for use with the MDR method that is a generalization of the normal-like-or-better criterion given in Troutt, Gribbin, Shanker and Zhang (2000) and discussed further in this type of validation might be based on the notion of unbiasedness. Thus, a test such as that based on the Hotelling T^2 statistic (see, for example, Morrison, 1976) might be considered for the hypothesis that the difference vectors, $\mathbf{z_t} = \mathbf{X}_t^o - \mathbf{X}_t^*$, have centroid, $\mathbf{0} \in \Re^k$. However, there are two difficulties with this strategy. First, some components of these vectors may not be ratio level measurements as in the example APP model of the chapter Appendix for which one variable is binary. Thus, the multivariate normal assumption for the Hotelling test cannot be met. Second, even if all the variables are ratio level measurements, we can expect the $\mathbf{X}_t^*$ to be extreme points or boundary points of their feasible regions, S_t, particularly when the stipulated model is linear or not strictly convex as in the model in the chapter Appendix. In that case, it may not be possible for the $\mathbf{X}_t^o$ to be distributed in an unbiased way around the $\mathbf{X}_t^*$, or equivalently, for the z_t to have centroid $\mathbf{0}$. That is, if the mode is on the boundary of S_t then again the distribution would not be expected to fit or approximate the multivariate normal law. For these reasons, we consider whether there is evidence that the mode, rather than the centroid, is $\mathbf{0}$ and propose a benchmark based on the observed δ-distribution that we call the target-mode agreement criterion.

When the distribution of $\mathbf{z}_t$ is unimodal with mode $\mathbf{0}$, and also it is spherically symmetric about $\mathbf{0}$, then the aim of unbiasedness may be regarded as establishing that the mode is $\mathbf{0}$. In that case, the mode and centroid are equivalent, as for example, when the $\mathbf{z}_t$ have a multivariate normal distribution. However, if the constraints geometry may not permit this symmetry, then we seek a different approach for assessing the plausibility that the mode is $\mathbf{0}$. Therefore, the target-mode agreement criterion has essentially the same purpose as establishing unbiasedness. To formalize this relationship we obtain a theorem connecting the mode of $V(\mathbf{x})$ to that of $f(\mathbf{x})$ based on Theorem 6.4 of Chapter 6.

Theorem 7.2 *Suppose $V(\mathbf{x})$ is a unimodal function on $\Re^k$ with minimum of zero at $\mathbf{x} = \mathbf{0}$. Let $f(\mathbf{x})$ be the pdf associated with $V(\mathbf{x}), g(u)$ and $\phi(u)$, where $\phi(u) = g(u)/A'(u)$. Assume that $A(u)$ is twice differentiable with $A'(u) > 0$. Let $g(u)$ be differentiable and assume that $g(u) > 0$ on $(0, \infty)$. Then a sufficient condition that the mode of $f(\mathbf{x})$ is $\mathbf{0}$ is that*

$$\frac{g'(u)}{g(u)} - \frac{A''(u)}{A'(u)} < 0, \quad \text{for } u > 0. \tag{7.10}$$

Proof. This result is essentially just a restatement of Theorem 6.4 in the case that the range of $V(x)$ is $[0, \infty)$. Details are omitted.

This result may be applied as follows. Given the estimated δ-values, interpreted as distances from a target at $\mathbf{0} \in \Re^k$, and a pdf, $g(\delta)$, with acceptable fit to the estimated δ-values, we argue that the target at $\mathbf{0}$ assumption is well supported if the mode of the corresponding $f(\mathbf{x})$ is $\mathbf{0}$. We call this the target-mode agreement (TMA) criterion. This can be reduced to a condition on the dimension, k, and the pdf, $g(\delta)$, using (7.10).

Let us employ the transformation $u = \delta^2$. Let $V(\mathbf{x}) = \mathbf{x}'\mathbf{x}$. The set $\{\mathbf{x} \in \Re^k : V(\mathbf{x}) \le \delta^2 = u\}$ has volume, $A(\delta^2) = A(u)$, given by $A(u) = \alpha_k u^{\frac{k}{2}}$, where $\alpha_k = \pi^{\frac{k}{2}}(\frac{k}{2}\Gamma(\frac{k}{2}))^{-1}$. See, for example, Fleming (1977). Therefore, $A'(u) = \frac{k}{2}\alpha_k u^{\frac{k}{2}-1}$, $A''(u) = \frac{k}{2}(\frac{k}{2}-1)\alpha_k u^{\frac{k}{2}-2}$ and $\frac{A''(u)}{A'(u)} = (\frac{k}{2}-1)u^{-1}$. From (7.10), the requirement for $g(u)$ to satisfy the target-mode agreement criterion is

$$\frac{g'(u)}{g(u)} < \left(\frac{k}{2} - 1\right) u^{-1}, \quad \text{for} \quad u > 0. \tag{7.11}$$

Example 7.1   For the APP model of the chapter Appendix with quarterly data, we have $N = 4$ and $p = 8$ so that $k = 32$. (In Section 7.4, we argue that a smaller value of k may be more appropriate.) Suppose that for a given data set, the estimated $u_t = \delta^2$ fit the gamma pdf, $g(u) = \beta^{-\alpha}[\Gamma(\alpha)]^{-1}u^{\alpha-1}\exp(-\frac{u}{\beta})$, for particular parameter values α_o and β_o. (See Law and Kelton, 1982, for a discussion of the gamma pdf and Madansky, 1988, for tests of fit.) For this pdf, $\frac{g'(u)}{g(u)} = (\alpha - 1)u^{-1} - \beta^{-1}$. Hence, (7.11) requires that $(\alpha_o - 1)u^{-1} - \beta_o - 1 < 13u^{-1}$ for $u \ge 0$. Thus, the condition requires $\alpha_o \le 16$. If $\alpha_o = 16$ (i.e. $\alpha_o = \frac{k}{2}$) and $\beta_o = 2$ then it can be verified using Theorem 1.2 that the corresponding $f(\mathbf{x})$ is the multivariate normal pdf. In this case, the performance of past attempts at achieving optimal plans according to the stipulated model can be said to have been as good as those governed by a multivariate normal distribution centered on a target at $\mathbf{0}$. If $\beta_o < 2 (\beta_o > 2)$ then $f(\mathbf{x})$ is steeper (flatter) than the multivariate normal pdf but is still finite at its mode. This is discussed further in Chapter 8 in connection with what we call the normal-like-or-better criterion. If $\alpha_o < 16$ then $f(\mathbf{x})$ is unbounded at its mode for every $\beta_o > 0$. Thus, for the gamma pdf, the TMA criterion reduces to $\alpha \le \frac{k}{2}$. However, it should be noted that the gamma pdf assumption for the u_t is not a necessary requirement. For instance, it might be found for a particular application, that the u_t fit the Weibull or lognormal, say, better than the gamma pdf. That is, Theorem 7.2 can be applied more generally.

Thus to summarize, if condition (7.10) is met, then the $g(\delta^2)$ pdf with $V(\mathbf{x}) = \mathbf{x}'\mathbf{x} = \delta^2$, corresponds to a pdf, $f(\mathbf{x})$, that has mode, $\mathbf{0}$. We propose this benchmark as acceptable evidence that the δ_t (or u_t) have been focused on the target of minimizing $V(\mathbf{x})$. This is an indirect approach. If the condition holds, then it is assumed that, the δ_t also represent acceptable performance in minimizing the stipulated planning model. In reality, the distribution of the $\mathbf{X}_t^o$ might be modeled according to pdf, $f_t(\mathbf{X})$, say, on S^t, preferably in such a way that the mode of $f_t(\mathbf{X})$ is $\mathbf{X}_t^*$. In that case, we have $V_t(\mathbf{X}) = E_t(\boldsymbol{\xi}^*; \mathbf{X}) - E_t(\boldsymbol{\xi}^*; \mathbf{X}_t^*)$, so that we may define $A_t(u) = L[\mathbf{X} \in S^t : V_t(\mathbf{X}) \leq u]$. Then the $f_t(\mathbf{X})$ might be specified by applying (7.10). However, because of the geometric complexity of the S_t and $V_t(\mathbf{X})$, and the possibility of alternative optima, it may not be practicable to obtain the $A_t(u)$ for realistic stipulated models like that in the chapter Appendix. Thus, we use the analytically tractable $V(\mathbf{X}) = \mathbf{X}'\mathbf{X}$ as a surrogate for the functions, $V_t(\mathbf{X})$, but assume that the u_t are independently and identically distributed with pdf $g(u)$.

7.4.2 *Stipulated Model Choice and Routine Validation*

The stipulated APP model has been taken as given for the previous discussion. Any model that is linear in the costs and for which the associated driver terms can be computed will be potentially suitable for the method. For example, the model in the chapter Appendix could be modified to include a cost term proportional to $(W_{t+h} - W_{t+h-1})^2$. This cost driver is one that penalizes change or lack of smoothness in the workforce size. Along with a similar term for inventory smoothness, this special type of quadratic cost term was the basis for the *Linear Decision Rule* theory. See Holt, Modigliani and Simon (1955) and Holt, Modigliani and Muth (1956).

We should emphasize that the proposed method does not reduce the need for routine validation of the stipulated planning model. Namely, given the cost parameter estimates and planning model that were actually implemented then on-going comparisons should be made between the total costs predicted, by the model and total costs actually realized from accounting data. Such comparisons may result in revisions of the known costs, c_i, $i \in I_o$, or the model form itself. Given such revisions, the proposed

MDR method can then estimate or re-estimate the costs, c_i, $i \notin I_o$, for which such validation may be difficult or impossible.

7.5 Data and Binary Variable Concerns

In this section, we discuss several data issues including data set size requirements, data availability and other possible applications. We also consider what can be called the *effective dimension* of the decision space and the impact of binary decision variables on the target-mode agreement criterion.

7.5.1 *Data Set Issues*

First, we consider the data set size, or number of observations, T, that may be required for the proposed estimation method. We note that the number of unknown costs may not be large. In the model of the Appendix, all costs except c_1, c_2, c_3, and c_8, are likely to be known or estimable with satisfactory precision. Thus, only four costs would need to be estimated by the MDR approach. Borrowing the 10k rule of thumb from regression analysis of a minimum of 10 observations for each unknown parameter suggests that $T = 40$ observations (past planning periods) should be sufficient. See Harrell (2001) and Kleinbaum *et al.* (1997). For example, three years of monthly data would approximately satisfy this criterion. In fact, we may argue that fewer than $T = 40$ observations should suffice, as the ξ_i are further known to be nonnegative and sum to unity. In addition, the known costs and bounds create additional constraints as discussed in Section 7.2. We next note that the $\mathbf{X}_t^o$ are assumed to have been feasible for problem, APP_t; so that $\mathbf{X}_t^o \in S^t$. This assumption might present difficulties due to data errors or changes in constraint parameters over the data periods. In the model of the Appendix, any or all of the parameters, $A_1 - A_{10}$, may have varied from period to period. For example, if the value of A_6, maximum inventory level permitted, had been decreased at some past period, then failure to reflect the new value in subsequent S^t constraint sets would possibly cause the associated data plans to be infeasible. Thus, care needs exerted to correctly specify S^t for each data period. In case infeasibilities remain, perhaps due to data errors, an adjustment is suggested below.

A data warehouse, if available, might be expected to provide some of the information needed for MDR models. However, actual planning data may not be routinely captured. Gray and Watson (1998) note that spreadsheet analyses, in particular, are seldom warehoused. In case past planning data are not available, then actually realized past demands might be substituted for missing past forecasts. Similarly, actual past decisions might be used to approximate plans. That is, the plan for period-t might be approximated by the actual decisions for the following N periods. However, such approximated actual plans must be feasible for the constraints in S^t. In case that does not hold for one or more t-values, then a reasonable expedient would be to select the plan in S^t that is nearest to the approximate plan. Assuming linear constraints as in the model of the chapter Appendix, this would require solving a mixed integer convex quadratic minimization program for each such case. That is, suppose for some t-value, the data vector $\mathbf{X}_t^o \notin S^t$. Then we propose as an adjusted data vector, the solution of the quadratic programming problem given by:

$$\text{Minimize:} \qquad (\mathbf{X} - \mathbf{X}_t^o)'(\mathbf{X} - \mathbf{X}_t^o), \qquad \text{subject to} \quad \mathbf{X} \in S^t. \qquad (7.12)$$

As noted in the Dubois and Oliff (1991) study, lack of cost information can be a barrier to the use of management science models. This point has been mentioned recently in Menon and Sharda (1999) and in Robinson and Dilts (1999) in connection with enterprise resource planning (ERP) systems and data mining. Many models that have been proposed assume that costs or similar parameters will be readily available for practitioners. Assistance with obtaining such information has been a generally neglected area. This chapter shows that the MDR approach can be applied to estimate such costs. ERP systems and associated data warehousing/data mining capabilities are now becoming widely deployed with interest for ORMS (Sohdi, 1999). Increasingly, these systems are adding planning and scheduling modules (Robinson and Dilts, 1999). Cost estimation can be expected to enable additional planning models to be included in such modules.

7.5.2 *Other Potential Application Settings*

We have used the production-planning context to motivate the MDR approach for behavioral estimation of costs. However, the method is applicable, in principle, to all situations in which repetitive decisions can be modeled with mathematical programming using an objective function that is linear in costs or similar parameters. Other estimation situations of this kind are as follows:

Linear programming allocation models. For repetitive cost minimizing allocation applications, it may be desirable to estimate one or more costs based on actual decisions taken. The objective function is linear in the costs and therefore is estimable by the MDR technique when the constraints are known. For profit or contribution-maximizing models, it should be noted that decisional regret must be defined as the model-optimal objective function value less that based on actual past decisions.

Weights for multiple objective function models. The foregoing situations involved costs or profit contributions. However, the weights for combining individual objective functions into a single weighted objective function could also be discovered, in principle, by using the MDR approach. That is, let the individual objective functions of interest to a decision maker be $f_i(\mathbf{x})$ for $i = 1, \ldots, n$. Then the past decisions may be modeled as solutions to the problem: minimize $\sum_{i=1}^{n} \lambda_i f_i(\mathbf{x})$, subject to a constraint set, $\mathbf{x} \in S$, where $\sum_{i=1}^{n} \lambda_i = 1$ and $\lambda_i \geq 0$ for all $i = 1, \ldots, n$. Here, regardless of the nature of the $f_i(\mathbf{x})$, the objective function is linear in the λ_i and is therefore suitable for the MDR approach provided that the constrained global minima of the $\sum_{i=1}^{n} \lambda_i f_i(\mathbf{x})$ can be computed.

Customer waiting costs in queues. In the economic analysis of waiting lines, there are two kind of costs. (See, for example, Stevenson, 1982). Costs for additional service capacity are usually easily specified in terms of machine or labor hour rates. However, customer waiting costs are much more difficult to estimate in an objective manner. An MDR approach should enable management to see what such costs are being implied by existing policies and then change the policies if necessary.

Cost of units short in single-period inventory problem. As noted earlier, Stevenson (1982) discusses an example of retrospective optimization for this problem. It appears that such problems are easily converted to MDR problems when more than one observation is available.

While we have emphasized the linear case in this chapter, solving the MDR estimation problem given by (7.10-7.11) should be more generally applicable. As an example, in the model of Gilbert (2000), parameters include costs and multiplicative seasonal factors. Moreover, the objective function of that model is not linear in the cost parameters. Nevertheless, the MDR estimation technique could still be applied with a meta-heuristic approach such as genetic algorithms. In fact, the constraint set (7.11) is of probability simplex form,

$$P_n = \left\{ \boldsymbol{\xi} \in \Re^n : \sum_{i=1}^{n} \xi_i = 1, \quad \xi_i \geq 0, \quad \text{for all } i = 1, \ldots, n \right\}, \qquad (7.13)$$

for which special genetic algorithm operators have been proposed. See Bhattacharyya and Troutt (2000, 2001). For a genetic algorithm approach, $\sum_{t=1}^{T} \delta_t(\boldsymbol{\xi})$ might serve as the fitness function. The $\mathbf{X}_t^*(\boldsymbol{\xi})$ vectors would be obtained as by-products of the fitness function evaluations, so that the genetic search is only required over the set P_n.

7.5.3 *Decision Space and Binary Decision Variables*

In Section 7.3.1 above, we discussed an example based on the model in the chapter Appendix with $N = 4$ periods in the planning horizon. This yielded plans of dimension $k = 32$, so that the decision space is contained in $\Re^{32}$. In the case of the gamma assumption for the u_t, it was found that the criterion requires $\alpha \leq k/2$ where α is the shape parameter of the gamma pdf. As the gamma shape parameter becomes smaller, the mode moves closer to the origin. (See, for example, Law and Kelton, 1982). Smaller α-values suggest better performance in concentrating more probability near the perfectly accurate case of $u = 0$. Therefore, if a smaller value of k is more appropriate, then a smaller α-value will be required to meet the TMA criterion.

Each equality constraint on the decision variables or system variables relating to decision variables of the stipulated model effectively reduces the dimension of the decision space by unity. The model in the chapter Appendix has five equality constraints ($A2, A4, A5, A7$ and $A8$) and one binary variable for each period. Since there are eight decision variables for each period, we have essentially only two real variables, or degrees of freedom, and one binary variable for each of the planning periods. This suggests that the decision space should be regarded as a subset of $\Re^8 \times \{0,1\}^4$ rather than $\Re^{32}$.

Intuitively, it appears inappropriate to use one full real dimension for each binary variable and replace $\Re^8 \times \{0,1\}^4$ by $\Re^{12}$. We now give an argument that the decision space should, in fact, be considered as contained in $\Re^8$; that is, the extra dimensions for the binary variables should be disregarded.

Let the objective function of a stipulated model be expressed as $\psi(\mathbf{y}) +$

$\gamma(\mathbf{z})$, where $\mathbf{y} \in \Re^{\eta}$ is the vector of real decision variables and $\mathbf{z} \in [0, 1]^{\nu}$ is a vector of binary variables. Let us use S' as the notation for the constraint set. For application of the TMA criterion, we use a surrogate for the set,

$$A(u) = L[\{(\mathbf{y}, \ \mathbf{z}) \in S', (\mathbf{y} \in \Re^{\eta}) : \psi(\mathbf{y}) + \gamma(\mathbf{z}) \leq u\}].$$

The subset of $\Re^{\eta}$ described by $A(u)$ is the union of the sets, $\{\mathbf{y} \in \Re^{\eta} : \psi(\mathbf{y}) + \gamma(\mathbf{z}) \leq u\}$, where the union is over the finitely many (2^{ν}) values of $\gamma(\mathbf{z})$. In any case, this set is a subset of $\Re^{\eta}$. Thus, we suggest that the appropriate value of k should be the effective dimension, defined to be η, the dimension of the real decision variables in the model. For the example of Section 7.3.1, we therefore obtain $k = 8$, for which the TMA criterion requires $\alpha \leq 4$. This is therefore a more strict requirement than the earlier $\alpha \leq 32$.

7.6   Re-estimation and Continuous Improvement

Here, we consider the question of how often it may be useful to repeat the estimation process. It may seem that once a stipulated model and its costs are satisfactorily identified then they could be used indefinitely thereafter. However, several possible events could affect the choice of models and costs and the need for re-estimation. First, as time progresses, some earlier, perhaps approximated, data may be augmented or replaced with more precise past planning data. Next, although certain costs were described as known in Section 7.2.1, it can be noted that most such costs are themselves estimates, which are subject to revision over time. As discussed in that section, the costs assumed to be known create constraints in the estimation model. Therefore, a revision of any known cost may have an impact on the estimates of unknown ones. Finally, as more data become available, it would also become possible to consider models that are more detailed. For example, even though a linear model may have been use in the past, it may later be of interest to consider a model that is quadratic in hiring costs, say, as in the chapter Appendix model. The MDR method provides a way to estimate the new linear and quadratic costs that would replace the old single linear one. Thus, re-estimation could help with continuous improvement in the model and parameter estimates.

7.7 Comparisons with Management Coefficients Theory

One of the central influences leading to the MDR principle was the Management Coefficients Theory (MCT) method proposed by Bowman (1963). See also Kunreuther (1969) and Moskowitz and Miller (1975). The two methods have different aims, however. MCT uses no cost considerations and seeks to reduce the variability of management decisions. The proposed method seeks to estimate cost parameters. Both methods should be useful in a system for monitoring production decisions and planning models. In addition, both methods fit actual management decision data to a model and are instances of what we have called behavioural estimation. In the MDR approach, the model is a specific planning model that depends on costs. MCT regresses actual decisions on the planning data.

MCT can also be described as an estimation technique for parameters of the Linear Decision Rule (LDR) model of Holt, Modigliani and Simon (1955) and Holt, Modigliani and Muth (1956). These latter two contributions can be called the HMMS theory and appear to have motivated Bowman's work.

In MCT, there were two regression models, one for workforce level changes and one for production quantities. These regression models used actual past decisions as dependent variables. The parameters obtained, called management coefficients, define decision rules that could be applied to current planning information to suggest new planning decisions. MCT assumes that management decisions are unbiased; that is, they are "accurate" on average. The MDR method assumes that management decisions are approximately optimal for a cost-minimizing model. If the "accurate" description is interpreted to mean optimal in MCT, then the two methods agree on this point. MCT regresses actual decisions on the planning data. Thus, the MCT approach begs the question of how to find optimal decisions that are near, in some sense, to the observed ones. The MDR approach provides an answer to this question.

MCT may be considered as a somewhat radical approach to achieving consistent management decisions. It is radical in the sense that it makes no references to costs. Bowman showed that cost performance would have been better under perfect and moving average forecasts in three of four company cases he examined, had the MCT results been used in place of actual past decisions. A study by Moskowitz and Miller (1975) gave strong additional support for the method. Bowman argued that the results provide evidence of the normative value of consistency in managerial decision-making. However, two important deficiencies of the method can be observed. First, management coefficients are abstract outside the context of MCT and are therefore difficult to compare and combine with known cost information. Second, regression results based on sub-optimal plans and decisions leave obvious room for improvement. Other deficiencies were pointed out by Eilon (1975), but reported tests of the method appear favorable, particularly when suitably modified (Kunreuther 1969).

One shortcoming of MCT is that the regression analyses should ideally be based on cost-minimizing decisions rather than actual decisions. Bowman apparently recognized the possibility of such a bias but did not attempt a detailed analysis or propose a method of elimination. He argued that the bias should be small and not critical due to the flatness of the convex expected cost functions near their minima. Lee and Khumawala (1974) noted that such a bias might explain the poorer performance of MCT in their study. However, other studies, as noted earlier, have found that the MCT performs well. The MDR approach explicitly finds optimal plans near to the actual ones and the difference between the identified model-optimal plans and the actual plans provides an estimate of the bias.

The proposed MDR method is not necessarily in competition with MCT as the two methods have different goals. However, consistency might also be achieved by adherence to a well-validated planning model. Let ξ^* be a solution of the MDR problem. Then total implicit planned costs using cost vector, ξ^*, and the stipulated planning model could only have been smaller than those using actual decisions with the same costs. This is because $\delta(\xi^*)$ is nonnegative for all t. Thus, if the forecasts were realized as planned then the firm would only have done better with model-optimal planning and the estimated cost vector. This is theoretically assured for the proposed method but was only demonstrated empirically for the MCT method.

7.8　Inappropriate Convergence

A potential practical case not considered earlier is that in which management has consistently used a fixed model, such as linear programming. Assuming the same model and the same costs for the MDR estimation, then that model could only recover exactly, what would already be known. This case is argued to be both unlikely to occur, as well as inappropriate from the viewpoint of sound practice. First, use of unmodified model recommendations is never advocated in the ORMS literature. Rather, it is always advised that model results be combined with other information and judgment in deciding what is to be implemented. In addition, APP decisions, particularly those involving workforce and inventory level changes, are likely to be the result of team decisions. Thus, judgmental modifications of any model's specific recommendations are routinely expected in actual practice.

It can be argued that such managerial model adjustments are actually useful, if not necessary, for gaining additional information. The present problem may be regarded as one of system identification. Permitting the identification process to converge to a particular model and its cost estimates would only be desirable if they were truly correct ones. Since that ideal state is unlikely to ever be verified exactly, some experimentation with the system seems advisable in order to prevent such *inappropriate convergence*. A similar idea, called the *dither signal*, is used in control theory (Brogan, 1991). Namely, expert drivers tend to test the brakes occasionally on wet or icy pavements in order to better estimate the appropriate braking technique or strategy that will be most effective. That is, otherwise optimal driving behavior is briefly modified in order to gain information that either validates or further improves subsequent driving behavior.

A related case might occur if management has used the same APP model, say, linear programming, but with different costs from period to period for a particular cost driver. If such cost changes cannot be regarded as *systematic* ones as discussed further in the next section, then we suggest that this indicates uncertainty about such costs, so that they cannot be regarded as among the known costs. A similar case might occur if constant cost parameters have been employed, but with different models at different points in the past. However, this case can also be considered as a way to reflect managerial judgment. For example, results from a planning approach based on a linear programming APP model in one or more previous planning periods might have yielded unacceptable inventory fluctuations. It could then be considered as a reasonable response by management to have employed a planning model with a quadratic inventory cost function in a subsequent period. Thus, changing to a new model might be interpreted as expressing judgment that other cost drivers should be given additional weight or consideration.

7.9   Miscellaneous Remarks

It is not known whether the choice of normalization, $\sum_{i=1}^{n} \xi_i = 1$, suggested for the MDR model is critical for the method. With little additional programming effort, a Euclidean norm constraint should be expected to work as well in all respects. Computational experiments might address that point. The objective function of the MDR model might also benefit from different weights for different periods to discount the influence of older data or for similar purposes.

Multiple alternative optima might occur for the retrospective optimal plans obtained in the MDR algorithm solutions. However, if the APP model is strictly convex then unique optimal solutions would be expected for these. Otherwise, selections of versions of these solutions could be problematical but potentially of interest in analyzing biases of the actual past decisions. If it were found that such biases tend to be nearly a constant vector, such information may be helpful in future planning. For example, analysis might indicate, say, a systematic tendency to use lower than optimal overtime levels. When the ξ_i of the MDR algorithm are further constrained by known cost relationships, it may be conjectured that cost estimates will tend to be uniquely optimal. Multiple alternative optima for the estimated cost vector might suggest a lack of independence or parsimony in choosing the cost drivers of the model.

We may also consider relationships among costs during the historical data period that may be called *systematic cost changes*. The MDR model, as developed above, implicitly assumed that all costs had either been stable or had risen or fallen according to some common index. This assumption may be adjusted in two special cases. First, suppose that a cost c_i has changed at a different rate from the common index of all the costs. Let ν_t be the desired index for this cost with $\nu_{t_o} = 1$. Then the appropriate cost for the period $t < t_o$ would be $\nu_t^{-1} c_i$, the adjustment needed to provide true relative values of the costs. The factor ν_t^{-1} may be considered to adjust the associated cost driver term for that period. Similarly, technological advances brought on line at some past period may have reduced a certain cost, such as that of regular time labor, by a fixed factor ζ. This is equivalent to regarding the prior costs to be ζ^{-1} times present levels. Adjusting the corresponding cost driver terms accordingly may reflect these factors.

We believe the MDR class of estimation methods will be more generally useful in application work and data mining (DM). MCT itself might be regarded as an early DM attempt. Many of the management science models that have been proposed assume the existence of cost parameters somewhat casually. Yet as indicated in the Dubois and Oliff (1991) study, such cost parameters are not necessarily easily obtained in the form needed for the models. Frequent reminders of this dependence on cost data appear in the practical literature of the field.

It may further be desirable to specify an average cost $\bar{c}$ for all cost components of a given type in a model which uses both linear and quadratic components. Such averages may be readily available from accounting data or special studies, as well as, informal managerial opinion. This case does not appear to be as readily incorporated into the MDR model. However, such data may be used for reasonableness or face validity assessment and adjustment of the model when at least one other cost is assumed to be known.

7.10   Conclusions

This chapter addresses a theory of behavioral estimation of costs and similar mathematical programming model parameters from actual decisions or actions. The approach is potentially applicable whenever similar decisions have been made repeatedly and a stipulated model for such decisions has an objective function, which is linear in the parameters to be estimated. The method was motivated by a production-planning model. In that setting, the method seeks to estimate those costs that are otherwise difficult to estimate. For well validated planning and estimation models, the method estimates such costs consistent with known costs and past management decisions. The main elements of the proposed method are the minimum decisional regret estimation principle and the target-mode agreement criterion.

Chapter Appendix - Aggregate Planning Example Model

This APP model expands on the linear programming model by including a second shift strategy involving zero-one variables and a convex quadratic cost component for hiring cost. The second shift modification uses the method of Glover and Woolsey (1974). We also provide for backorders. For simplicity, no sub-contracting is included. The notation and modeling strategy of Vollmann *et al.* (1995) are generally followed. In particular, the model is based on the model given in pages 622-624 of that source except that an average hours worked assumption was made for regular and second shift employees. Thus, idle time was not explicitly modeled in the interest of parsimony. Similarly, we do not specify ending workforce and inventory levels. This approach has production, demand, and inventory levels expressed in direct labor hours. Thus, the example attempts to convey generality without excessive detail. The second shift option may be regarded as a capacity change. For a general approach to capacity changes in production-planning, see Rajagopalan and Swaminathan (2001). Thus, define:

Costs:

$$c_1 = L_H = \text{the linear term coefficient for hiring cost.}$$

$$c_2 = Q_H = \text{the quadratic term coefficient for hiring cost.}$$

$$c_3 = C_F = \text{the cost of firing an employee.}$$

$$c_4 = C_P = \text{the cost per labor-hour of regular time production.}$$

$$c_5 = C_O = \text{the cost per labor-hour of overtime production.}$$

$$c_6 = C_Y = \text{the cost per labor-hour of second shift production.}$$

$$c_7 = C_I = \text{the cost per period of carrying one labor-hour of production.}$$

$$c_8 = C_B = \text{the cost per period per labor-hour backordered.}$$

$$c_9 = C_Z^+ = \text{the start-up cost for the second shift.}$$

$$c_{10} = C_Z^- = \text{the cost of closing the second shift.}$$

Decision Variables:

$$x_{1,t} = H_t = \text{the number to be hired in period-}t.$$

$$x_{2,t} = F_t = \text{the number fired in period-}t.$$

$$x_{3,t} = P_t = \text{regular time production hours scheduled in period-}t.$$

$$x_{4,t} = O_t = \text{the overtime production hours scheduled in period-}t.$$

$$x_{5,t} = Y_t = \text{second shift production hours scheduled in period-}t.$$

$$x_{6,t} = I_t^+ = \text{the ending inventory in labor-hours for period-}t.$$

$$x_{7,t} = I_t^- = \text{the labor-hours backordered at the end of period-}t.$$

$$x_{8,t} = Z_t \quad = \quad \text{a zero-one variable such that } Z = 0 \text{ indicates no second}$$
$$\text{shift is operating and } Z = 1 \text{ indicates that the second}$$
$$\text{shift is operating at the end of period-}t.$$

System variables:

$W_t \quad = \quad$ total workforce size at the end of period-t.

$S_t^+ \quad = \quad$ second shift workforce size at the end of period-t.

$S_t^- \quad = \quad$ amount by which the regular shift workforce size at the end of period-t falls short of its maximum.

$Q_t \quad = \quad Z_{t-1}Z_t$, a variable used in the method of Glover and Woolsey (1974).

Data and parameters:

$D_t \quad = \quad$ the forecasted labor-hours to be sold in period-t.

$A_1 \quad = \quad$ average number of regular time hours worked per employee per period.

$A_2 \quad = \quad$ average number of hours to be worked per second shift employee per period.

$A_3 \quad = \quad$ maximum number of overtime hours worked per employee per period.

$A_4 \quad = \quad$ maximum workforce size for the regular shift.

$A_5 \quad = \quad$ maximum workforce size for the second shift.

$A_6 \quad = \quad$ maximum inventory level permitted.

$A_7 \quad = \quad$ minimum second shift workforce size.

$A_8 \quad = \quad$ inventory initialization constants

$A_9 \quad = \quad$ inventory initialization constants

$A_{10} \quad = \quad$ initial workforce size.

$Z_0 \quad = \quad$ Second shift initialization constant (0 or 1).

$M \quad = \quad$ a large constant for use in a fixed-charge type integer constraint.

$N \quad = \quad$ number of periods in the planning horizon

Objective function:

$$\min \quad \sum_{h=1}^{N} \{ (L_H H_{t+h} + Q_H H_{t+h}^2 + C_F F_{t+h} + C_P P_{t+h} + C_O O_{t+h}$$
$$+ C_Y Y_{t+h} + C_I I_{t+h}^+ + C_B I_{t+h}^- + C_Z^- (Z_{t+h-1} - Q_{t+h})$$
$$+ C_Z^+ (Z_{t+h} - Q_{t+h}) \}$$

$$(A.1)$$

Subject to:

1. Inventory constraints:

$$I_{t+h-1}^+ - I_{t+h-1}^- - I_{t+h}^+ + I_{t+h}^-$$
$$+ P_{t+h} + O_{t+h} + Y_{t+h} = D_{t+h}, \quad h = 1, \dots, N$$

$$(A.2)$$

$$I_{t+h}^+ \leq A_6, \quad h = 1, \dots, N$$

$$(A.3)$$

2. Regular time production constraints (noting that $W_t - S_t^+ = A_4 - S_t^-$ is the regular shift workforce size - see also constraints A.8 below):

$$P_{t+h} - A_1 W_{t+h} + A_1 S_{t+h}^+ = 0, \quad h = 1, \dots, N$$

$$(A.4)$$

3. Second shift production constraints:

$$Y_{t+h} - A_2 S_{t+h}^+ = 0, \quad h = 1, \dots, N$$

$$(A.5)$$

4. Overtime production constraints, assuming overtime available only to first shift employees:

$$O_t + h - A_3 W_{t+h} + A_3 S_{t+h}^+ \leq 0, \quad h = 1, \dots, N$$

$$(A.6)$$

5. Workforce level change constraints:

$$W_{t+h} - W_{t+h-1} - H_{t+h} + F_{t+h} = 0, \quad h = 1, \dots, N$$

$$(A.7)$$

6. Workforce and shift relationships:

$$W_{t+h} - A_4 - S_{t+h}^+ + S_{t+h}^- = 0, \quad h = 1, \ldots, N$$

$$(A.8)$$

$$S_{t+h}^+ \leq A_5, \quad h = 1, \ldots, N$$

$$(A.9)$$

7. Relationship of second shift binary variables to second shift production levels:

$$Y_{t+h} - MZ_{t+h} \leq 0, \quad h = 1, \ldots, N$$

$$(A.10)$$

8. Minimum allowable second shift workforce size

$$S_{t+h}^+ \geq A_7 Z_t$$

$$(A.11)$$

9. Linear reduction of the product variable, $Q_t = Z_{t-1} Z_t$ using the method of Glover and Woolsey (1974):

$$Q_{t+h} - Z_{t+h} \leq 0, \quad h = 1, \ldots, N \qquad (A.12)$$
$$Q_{t+h} - Z_{t+h-1} \leq 0, \quad h = 1, \ldots, N \qquad (A.13)$$
$$Z_{t+h-1} + Z_{t+h} - Q_{t+h} \leq 1, \quad h = 1, \ldots, N \qquad (A.14)$$

Using (A.12) - (A.13), $0 \leq Q_{t+h} \leq 1$, and the Z_{t+h} as binary variables, the last two terms of (A.1) can be checked to give the shift change costs.

10. Initialization constraints:

$$I_t^+ = A_8 \qquad I_t^- = A_9 \quad W_t = A_{10} \qquad Z_t = Z_0 \qquad (A.15)$$

11. Variable ranges constraints:

$$H_{t+h}, F_{t+h}, P_{t+h}, O_{t+h}, Y_{t+h}, I_{t+h}^+ \geq 0$$

$$I_{t+h}^-, W_{t+h}, S_{t+h}^+, S_{t+h}^- \geq 0, \quad h = 1, \ldots, N$$

$$(A.16)$$

$$Z_{t+h} = 0 \text{ or } 1, \quad 0 \leq Q_{t+h} \leq 1, \quad h = 1, \ldots, N$$

$$(A.17)$$

The foregoing model is a mixed integer positive semi-definite quadratic programming model. A branch and bound solution strategy can be employed in which relaxed sub-problems are quadratic minimization problems. Because of the lack of strict convexity of the objective function (A.1), an appropriate quadratic programming algorithm should be employed. See Avriel (1976). Further details on this model and a solution example for simulated data are given in Troutt *et al.* (2001).

Chapter 8

Management Science Applications of VDR–III

Estimation of Benchmark Costs and Cost Matrices

In this chapter, we consider the activity-based costing (ABC) situation, in which for each of several comparable operational units, multiple cost drivers generate one or more distinct cost pools. In Sections 8.1.-8.6, we deal with the simpler case of a single cost pool. Here, we define what may be called benchmark or most efficient costs per unit of driver. A principle of *maximum performance efficiency* (MPE) is proposed and an approach to estimating the benchmark costs is derived from this principle. In Chapter 6 we introduced the NLOB effectiveness concept for scalar scores in the unit interval. Here, we develop a multivariate application of that concept. We illustrate the approach on published data from a set of property tax collection offices, called rates departments, for the London metropolitan area. Application to longitudinal data on a single unit is briefly discussed. We also consider some implications for the routine case when costs objects are disaggregated to sub-pools associated with individual cost drivers.

In the rest of the chapter, we consider the more complex situation in which there are multiple cost pools. We also propose a more sophisticated use of the MPE method. We define *technically efficient benchmark cost matrices*, which give costs of each category per unit of each cost driver. Such matrices may be regarded as technological coefficients for an auxiliary linear programming problem, which we call the *target model.* Our discussion focuses on academic departments within a college of business. In this context, cost drivers or outputs are the credit hours produced in various instructional activity categories or types. These drivers generate

179

two distinct cost pools. Then actual performance of the units is compared to the target linear programming model for which the cost matrix and objective coefficient vector are unknown parameters. These parameter sets are estimated using the MPE principle, which is a variant of the MDE principle, combined with the uniform conditional probability density function assumption to construct a maximum likelihood procedure. MPE was introduced in Troutt *et al.* (2000) and is similar to the MDE principle introduced earlier in Troutt (1995). In addition, both of these methods can be related to the MDR approach of Chapter 7. A combined mathematical programming and genetic search process is proposed for solving the maximum likelihood problem. This chapter illustrates the value of these estimation techniques for benchmark cost applications. Sections 8.1 to 8.7 follow results in Troutt *et al.* (2000) and in Troutt *et al.* (2003a).

8.1 The Multiple Driver - Single Cost Pool Case

We discuss an area of activity-based costing (ABC) that has received little attention in the literature. In what may be called routine ABC, the general procedure for obtaining unit cost rates is as follows (Horngren, Foster, and Datar, 2000). Costs for an activity are accumulated in "cost pools". A variable, called a *cost driver*, is identified that measures the amount or extent of the activity performed and that, ideally, varies proportionally with the cost pool level. Then the cost per unit of driver, sometimes called a unit cost, is found by dividing the cost pool's dollar amount by the associated driver level. The respective cost drivers are generally used as the basis for allocation of costs to specific cost objects (e.g., products or services). This has frequently been characterized by observing that products or services generate activities, which in turn generate costs. This procedure requires that costs can be disaggregated in such a way as to be associated to a single cost driver. In practice, it is quite often possible to carry out ABC in this way by sufficient information refinement. In some cases, however, more than one cost driver may simultaneously impact or "drive" a cost pool. The data set due to Dyson and Thanassoulis (DT) (1988), and Thanassoulis, Dyson and Foster (TDF) (1987), provides an example and is discussed below. In addition, in the comprehensive study of the airline industry by Banker and Johnston (1993), two or more cost drivers were found to drive the major cost pools with high statistical significance.

As shown in this chapter, the multiple cost driver case can be incorporated into a cost efficiency model useful for benchmarking comparable operational units in a firm, industry or other grouping. We illustrate a model-based method for benchmarking comparable operational units with multiple cost drivers. The method yields relative *cost efficiencies* of the units and provides estimates of what we call benchmark costs.

Suppose that x is a cost pool amount associated with the two simultaneous cost driver levels, y_1 and y_2. We wish to determine what unit cost rates or costs per unit, called simply costs when the meaning is clear, should be associated with these two drivers. Let a_1 and a_2 be the cost rates per unit of y_1 and y_2, respectively. Then under the assumptions of linearity and constant returns to scale of the total cost function, these costs must satisfy the total cost equation $x = a_1y_1 + a_2y_2$. Of course, there exists no unique solution in this single observation case. However, if several observations of x_j, y_{1j} and y_{2j} are available then modeling possibilities exist for estimation of the costs. In particular, regression through the origin with the model

$$x_j = a_1y_{1j} + a_2y_{2j} + \varepsilon_j \tag{8.1}$$

may be used to estimate what DT called "average costs". This type of regression model was also the basis of the approach used in Banker and Johnston (1993). Thus, if data for several comparable units, or several observations of the same unit over time, are available, then regression through the origin may be used to estimate costs (average unit cost rates) in the multiple simultaneous cost driver case.

However, there is a difficulty in the regression approach when the goal is to compare the units for what we call *cost efficiency*. That is, for benchmarking the efficient cost performance of the units, it is desired to estimate the cost rates of the most efficient unit(s). The specific difficulty is that for one or more of the units, the x_j value may be larger than necessary for the associated y_{1j} and y_{2j} due to cost inefficiency. Namely, let a_1^o and a_2^o be the costs for the most efficient unit(s). We call these the *benchmark costs*. Thus, in general, we must have

$$a_1^o y_{1j} + a_2^o y_{2j} = x_j^o \leq x_j \qquad \text{for all } j, \tag{8.2}$$

where x_j^o is the (unobserved) total cost associated with full efficiency, had it been achieved by unit-j. Furthermore, if a_{1j} and a_{2j} are the actual cost rates for unit-j (also unobserved) then the total cost function for unit-j is

$$a_{1j}y_{1j} + a_{2j}y_{2j} = x_j, \qquad (8.3)$$

so that the inefficiency of unit j , if any, may be decomposed as

$$s_j = x_j - x_j^o = (a_{1j} - a_1^o)y_{1j} + (a_{2j} - a_2^o)y_{2j} \qquad (8.4)$$

This decomposition suggests that it is possible to attain both benchmark costs simultaneously. In addition, a measure of cost efficiency for the j-th unit is given by

$$v_j = x_j^o/x_j \qquad (8.5)$$

Hence, inefficiency is assumed to occur for such a unit when it has one or more unit cost rates larger than the corresponding benchmark cost rates.

8.1.1   *The Importance of Cost Benchmarking*

Firms have frequently formed voluntary cooperative arrangements to share benchmarking information (Elnathan and Kim, 1995, Elnathan *et al.*, 1996). Some have collaborated (at an expense to their individual firms) and hired a management-consulting firm to study their costs or have used databases of external management consulting firms for benchmarking purposes. Many organizations that use benchmarking may incur significant related costs. While there may be significant costs associated with benchmarking, there are also many benefits. Elnathan and Kim (1995) discuss three sources of changes in profits that may accrue when firms collaborate in cooperative benchmarking. First, firm profits may increase due to improvements in operations (increased productivity or reduced production costs). Second, a firm's competitive advantage within its industry may change firm profits due to information sharing. Third, there may be other political, social, or control-related effects of benchmarking. Even though there may be significant costs associated with benchmarking, firms undertake benchmarking efforts because they view the benefits to be gained as outweighing the costs. For further discussion on this issue, the reader is referred to Troutt *et al.* (2000) or Troutt *et al.* (2003b).

Ideally, the benchmarking process should: 1) identify organizational units whose practices or procedures can be improved in terms of efficiency; 2) identify outstanding performers for emulation; 3) provide an operational measure of efficiency; and 4) provide performance targets in terms of unit costs of activities or similar figures to which management can relate. The present technique substantially accomplishes these ends.

8.1.2 *The Rates Departments Data*

The reader is referred to the DT and TDF papers for a more detailed description. These data were collected for a set of 62 property tax collection offices, called rates departments, in the London Boroughs and Metropolitan Districts. Total annual costs, measured in units of £ 100,000, for these offices (units), were collected along with activity cost driver levels, called outputs in DT and TDF, for four activities. The first three activities, collection of non-council hereditaments, rates rebates generated, and summonses issued and distress warrants obtained, were measured in units of £ 10,000, £ 1,000, and £ 1,000, respectively. The fourth, net present value of non-council rates collected, was measured in units of £ 10,000. This last one was included to reflect the additional administrative effort exerted to ensure the timely payment of large accounts. Thus, this data set gives total costs and cost drivers for four activities. We wish to determine the cost efficiencies of the units and estimate the activity units costs for the most efficient unit(s).

8.2   Other Modeling Approaches

The "routine ABC" method referred to above appears to be the basis of most previous benchmarking studies. Two model-oriented approaches were found in the literature for benchmarking analysis. One of these is due to DT and was based on modified Data Envelopment Analysis (DEA, Charnes *et al.*, 1994). Dopuch and Gupta (1997) recently proposed the use of stochastic frontier estimation (SFE) for estimating benchmark standards in a public education setting. Both of these approaches are reviewed in this section. Further discussion of the routine ABC approach is given below.

TDF and DT considered the application of DEA to the data set considered here. The weights estimated in DEA models correspond to the cost estimates needed in this study. A shortcoming of DEA, from the viewpoint of this chapter's goal, is that the particular weights that render one unit technically efficient (Pareto-Koopmans efficient) may differ from those for another unit. This is called *weights flexibility* by DT and is discussed further below. In addition, a relatively large number of units are typically declared fully efficient. Thus, no consensus on weights is achieved in DEA. DT proposed a heuristic modification of DEA. They first estimated average costs by regression through the origin as discussed above. Then, half of the respective average costs were used as reasonable lower bounds for values of the weights, respectively. While this approach reduces the weights flexibility, it is not removed entirely. Also, it may be argued, especially for benchmarking, that obtaining reasonable lower bounds on the weights (benchmark costs) is a central part of the problem. From that perspective, the DT method may be considered as somewhat *ad hoc*, even if reasonable. However, primary interest in that work was computation of efficiency scores with weights that do not vary as widely as those in unmodified DEA. Here, we seek consensus on the benchmark cost estimates as well.

Dopuch and Gupta (1997) proposed a benchmarking model using a Stochastic Frontier Estimation (SFE) method due to Aigner, Lovell and Schmidt (1977). They applied their model to evaluating the cost efficiency for a segment of the Missouri public school system in a data set similar to the one used here. In general, SFE models first define a parametric frontier model, which represents best possible performance, minimum or maximum, depending on context. Then actual performance is modeled as the frontier model plus an error term composed of two parts. The first error term is assumed to be normally distributed with mean zero. It is usually regarded as accounting for uncertainty in the frontier model. The second error term is a nonnegative one representing a measure of inefficiency or deviation from the efficient frontier. The Aigner, Lovell and Schmidt (1977) method assumes that such nonnegative inefficiencies are distributed as half-normal. The method proposed here does not require a preliminary assumption on the form of this density.

Other SFE approaches have also been proposed but do not appear to have been applied to cost efficiency as defined here. Green (1990) considers a model that assumes a gamma distribution for the inefficiency error terms. However, Ritter and Léopold (1997) have found that such models are difficult to accurately estimate. Recently, van den Broeck *et al.* (1994) have also considered Bayesian SFE models. We leave Bayesian considerations beyond the scope of the present discussion.

8.3 Model Development

Suppose we have $j = 1, \ldots, N$ *comparable* business units, achieving y_{rj} units of driver $r = 1, \ldots, R$, respectively, and with associated cost pools, x_j. In addition to having the same activities and cost drivers, we further require that comparable units be similar in the sense that the practices, policies, technologies, employee competence levels and managerial actions of any one should be transferable, in principle, to any other. Define a_r^o as the cost rates associated with the most efficient unit or units under comparison. Then, in

$$\sum_{r=1}^{R} a_r^o y_{rj} + s_j = x_j \qquad j = 1, \ldots, N, \tag{8.6}$$

the term s_j, may be interpreted as an inefficiency error component as in the SFE models. The ratio $v_j = \sum_{r=1}^{R} a_r^o y_{rj}/x_j$, is an efficiency measure for the j-th unit when the a_r^o are the true benchmark cost rates and $\sum_{r=1}^{R} a_r^o y_{rj} \leq x_j$ holds for all units. We call v_j the cost efficiency of unit-j.

A technique for estimating parameters in efficiency ratios of the above form was proposed in Troutt (1995). In that paper, the primary data were specific values of decision variables. Here a more general form of that approach called *maximum performance efficiency* (MPE) is proposed and applied to estimate the benchmark a_r-values. Assume that each unit $j = 1, \ldots, N$ seeks to achieve maximum (100%) efficiency. Then the whole set of units may be regarded as attempting to maximize the sum of these efficiency ratios, namely, $\sum \sum a_r^o y_{rj}/x_j$.

The maximum performance efficiency estimation principle proposes estimates of the a_r^o as those that render the total, or equivalently the average, of these efficiencies a maximum. This estimation criterion is a variation of the maximum decisional efficiency (MDE) principle (Troutt, 1995). The MPE approach is technically the same as MDE but is applied to general performance measures rather than values of decision variables. The MDE principle assumes that decisions are made, i.e. decision values are realized, in such a way as to maximize average efficiency relative to the model in question. MPE is a restatement in that performance vector values are realized in such a way as to maximize average efficiency relative to the model in question again, but where the model is expressed in terms of the performance vectors. It may be stated as follows:

Maximum Performance Efficiency Principle: *In a performance model depending on an unknown parameter vector, select as the estimate of the parameter vector, that vector for which the average performance efficiency ratio is greatest.*

Let $Y_{rj} = y_{rj}/x_j$, where y_{rj} is the driver level (output value in DT) for the r-th activity in operational unit-j and x_j is the total cost for unit-j. For the rates departments data, we have $N = 62$ operational units and $R = 4$ cost drivers. The Y_{rj} are the performance vectors to which the MPE estimation method applies. Then the estimation model for the benchmark a_r^o-values is given by

MPE:

$$\max \sum_{j=1}^{N} \sum_{r=1}^{R} a_r Y_{rj} \tag{8.7}$$

$$\text{s. t.} \quad \sum_{r=1}^{R} a_r Y_{rj} \leq 1, \quad \text{for all } j, a_r \geq 0 \quad \text{for all } r \tag{8.8}$$

The MPE model is a linear programming (LP) problem whose unknown variables are the benchmark unit costs, the a_r^o-values. Solution of this model provides values for the a_r^o, as well as the unit efficiencies, v_j. The model was applied to the rates departments' data using the LP option of the Solver Tool in Microsoft Excel^{TM}. The resulting estimates were:

$$a_1^o = 0.00000, a_2^o = 0.08820, a_3^o = 0.2671, a_4^o = 0.0664 \tag{8.9}$$

Unfortunately, there are two concerns about this raw MPE model solution. First, as part of their analysis, DT obtained an estimate of the average costs by regression through the origin as described above. Their results were

$$\bar{a}_1 = 0.05042, \bar{a}_2 = 0.07845, \bar{a}_3 = 0.1765, \bar{a}_4 = 0.1940 \tag{8.10}$$

Clearly, minimum or benchmark activity unit cost estimates should not exceed average ones. However, it can be noted that the estimates for both

a_2^o and a_3^o from the raw MPE model exceed the corresponding average cost estimates from the DT regression model.

A second solution quality issue is the zero value estimated for a_1^o. For the present, we consider an estimated benchmark cost of zero for an activity to be unreasonable. In particular, activity-one is a major activity that does, in fact, have positive average costs. While it may be conceivable that some unit can achieve one or more activities as cost-free by-products of one or more others, we believe that conclusive identification of such cases requires further research.

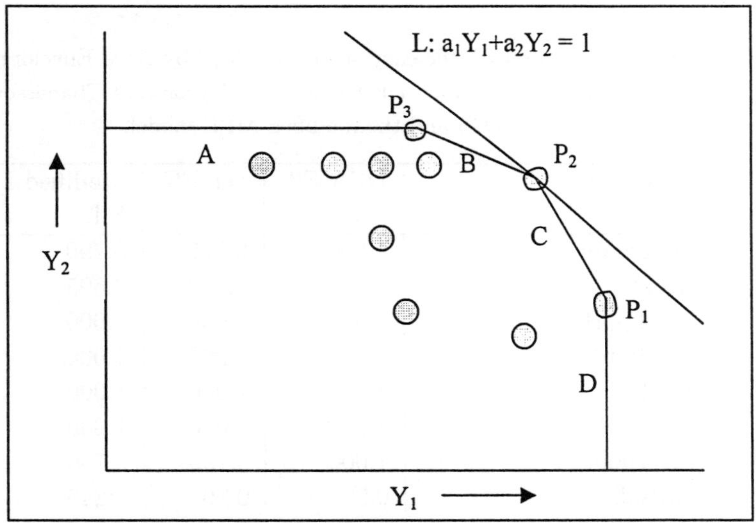

Figure 8.1: Hypothetical two-dimensional analog
of the rates departments data

Thus, it is necessary to modify the raw MPE model to improve the solution quality of its estimates. To understand the modification, consider Figure 8.1, which depicts a two-dimensional version of the present data set but uses hypothetical data points. This figure indicates the efficient frontier formed by the three indicated DEA efficient units with data vectors at points P_1, P_2 and P_3. The linear functions defining the efficient frontier facets A-D, respectively, correspond to the candidate basic feasible solutions of the raw MPE model. The solution associated with the frontier segment D is the raw MPE solution. This would be a vertical line with $a_2^o = 0$, namely a zero parameter estimate. In addition, this solution would assign only unit P_1 as having full efficiency. However, if the model is constrained to require that the unit at P_2 be efficient, then Line L depicts feasible values of a_1^o and a_2^o for the MPE model. The set of such feasible values has extreme points associated with rotation of L until it is coincident with frontier segments B or C, respectively. Both these segments have negative slope and therefore both parameter estimates will be positive. Thus, the modified MPE model is as follows. A separate run of the MPE model was made for each DEA-efficient unit, requiring that unit to be fully efficient. The seven DEA-efficient units were previously identified in DT and are shown in Table 8.1, Rows 1-7.

Table 8.1: Comparison of efficiency scores obtained by Data Envelopment Analysis (DEA), the modified DEA method of Dyson and Thanassoulis (1988)(DT) and the modified MPE model.

No.	Rates Dept.	DEA Eff.	DT Eff.	Modified MPE Eff.
1	Lewisham	1.000	0.827	0.790
2	Brent	1.000	0.743	0.695
3	Stockport	1.000	1.000	1.000
4	Bradford	1.000	0.999	1.000
5	Leeds	1.000	1.000	1.000
6	City of London	1.000	1.000	1.000
7	Liverpool	1.000	0.796	0.760
8	Walsall	0.996	0.861	0.840
9	Rotherham	0.994	0.849	0.795
10	Wakefield	0.993	0.890	0.866
11	Lambeth	0.961	0.834	0.816
12	Sunderland	0.942	0.801	0.753
13	Solihull	0.931	0.917	0.899
14	Redbridge	0.847	0.827	0.814
15	Calderdale	0.842	0.818	0.802
16	Haringey	0.822	0.710	0.690
17	Barking and Dagenham	0.801	0.644	0.610
18	Newcastle-upon-Tyne	0.798	0.713	0.703
19	Manchester	0.789	0.641	0.626

Table 8.1: Continued

No.	Rates Dept.	DEA Eff.	DT Eff.	Modified MPE Eff.
20	Wolverhampton	0.782	0.686	0.667
21	Trafford	0.761	0.756	0.751
22	Tameside	0.759	0.705	0.683
23	St. Helens	0.757	0.694	0.670
24	Sutton	0.746	0.692	0.659
25	Rochdale	0.745	0.718	0.696
26	Barnsley	0.714	0.617	0.599
27	Kirdlees	0.713	0.697	0.690
28	Oldham	0.702	0.687	0.679
29	Sheffield	0.702	0.702	0.695
30	Havering	0.700	0.698	0.695
31	Dudley	0.700	0.672	0.659
32	Sefton	0.690	0.677	0.664
33	Bexley	0.688	0.682	0.669
34	Gateshead	0.686	0.621	0.605
35	Wigan	0.683	0.652	0.639
36	Kensington and Chelsea	0.676	0.587	0.570
37	Coventry	0.674	0.645	0.631
38	Sandwell	0.644	0.604	0.593
39	Bury	0.639	0.638	0.632
40	South Tyneside	0.635	0.526	0.483
41	Salford	0.629	0.590	0.581
42	Hackney	0.614	0.468	0.445
43	Camden	0.597	0.562	0.556
44	Hillingdon	0.588	0.587	0.587
45	Tower Hamlets	0.568	0.529	0.523
46	Barnet	0.568	0.567	0.563
47	Bolton	0.557	0.549	0.543
48	Ealing	0.556	0.545	0.542
49	Bromley	0.548	0.520	0.506

Table 8.1: Continued

No.	Rates Dept.	DEA Eff.	DT Eff.	Modified MPE Eff.
50	Wandsworth	0.543	0.524	0.511
51	Birmingham	0.535	0.500	0.491
52	Enfield	0.516	0.512	0.505
53	Southwark	0.509	0.470	0.464
54	Knowsley	0.500	0.487	0.481
55	Islington	0.496	0.420	0.411
56	North Tyneside	0.465	0.465	0.461
57	Kingston-upon-Thames	0.442	0.426	0.413
58	Hounslow	0.435	0.433	0.430
59	Richmond-upon-Thames	0.431	0.410	0.396
60	Hammersmith and Fulham	0.424	0.373	0.364
61	Newham	0.333	0.331	0.329
62	Merton	0.329	0.302	0.286
	Means	0.705	0.652	0.637
	Standard Deviations	0.187	0.168	0.166

In each such step, it was only necessary to change the inequality constraint for that unit to an equality constraint. The best of these seven solutions was identified by having all positive cost estimates and otherwise the maximal value of the objective function. Thus, the modified MPE model preemptively requires a maximal number of positive cost estimates. The efficiency values obtained are shown in Table 8.1, Column 5. The corresponding cost estimates were:

$$a_1^o = 0.261807, \ a_2^o = 0.049353, \ a_3^o = 0.139833, \ a_4^o = 0.127998 \quad (8.11)$$

This solution passes the two tests of reasonableness above. Namely, all parameter estimates are positive; and no parameter value exceeds its average counterpart from the regression model. In addition, this model identifies four of the units as having maximal efficiency of unity. Four is the largest number of units that can be declared fully efficient since four linearly independent efficient points define a facet hyper-plane of the efficient frontier for the dimensions of this data set. We regard this as an additional measure of solution quality from the viewpoint of consensus. Four units that agree on the cost estimates is the best consensus achievable in this case. Thus, this solution has a number of desirable features. However, it is subjected to a further statistical test of reasonableness below.

Figure 8.1 also helps to highlight the differences between DEA and the proposed model. Application of DEA to this data set was fully discussed in TDF and DT with resulting DEA efficiency scores as shown in Table 8.1. The seven fully DEA-efficient units are in the first seven rows of the table. In terms of the hypothetical data set of Figure 8.1, DEA would identify the three fully DEA-efficient units at P_1, P_2 and P_3. That is, DEA identifies the efficient frontier extreme points. Optimal DEA output weights would also be obtained for each unit. These correspond to the cost estimates and describe the coefficients in the lines that define the four facets, A-D, respectively. The solution of the modified MPE model corresponds to facet C. The coefficients of the linear function defining that facet are both positive and otherwise maximize the average efficiency of all the units. Thus, DEA is useful prior to applying the present model in order to determine the efficient frontier extreme points.

In the MPE model, we have assumed only inefficiency error terms. This appears to be justified since we are able to specify the benchmark total cost function exactly, up to the benchmark cost parameters. Hence, there is no need to consider the additional normal error terms of the SFE approach.

It appears possible that there could exist data sets for which no solution of the modified MPE model yields all positive weights. For example, if the Y_{rj} data for some operational unit dominate all the other data then the efficient frontier boundary of Figure 8.1 might resemble a rectangle. It that case, we would accept the estimates obtained from the solution of the unmodified MPE model, provided such solution is unique. Another possibility would be to consider the dominating operational unit for removal as a possible outlier. Failing these possibilities, it may not be possible to obtain estimates of benchmark costs along the present lines.

8.4 Normal-Like-or-Better Performance

A basic assumption underlying the MPE estimation principle's applicability, as with the MDR principle, is that the sample of units under analysis have the goal of achieving maximum (100%) efficiency. This is a model aptness issue that parallels the requirement of $N(0, \sigma^2)$ residuals in OLS regression theory. In the present MPE case, the corresponding issue is to specify a distribution characteristic of the v_j that indicates consistency with a goal or target of unity (1.0) efficiency. In the previous chapter, we discussed the TMA criterion. The TMA criterion evolved from what we call the *normal-like-or-better* (NLOB) effectiveness criterion for these fitted efficiency scores. The NLOB criterion was first discussed in Troutt *et al.* (2000), and is described next.

As a model for appropriate concentration on a target we begin with an interpretation of the multivariate normal distribution, $N(\boldsymbol{\mu}, \boldsymbol{\Sigma})$, on $\Re^n$. If a distribution of attempts has the $N(\boldsymbol{\mu}, \boldsymbol{\Sigma})$ or even higher concentration of density at the mode $\boldsymbol{\mu}$, then we propose this as evidence that $\boldsymbol{\mu}$ is indeed a plausible target of the attempts. This is exemplified by considering a distribution model for the results of throwing darts at a bull's-eye target. Common experience suggests that a bivariate normal density represents such data reasonably well. Steeper or flatter densities would still be indicative of effective attempts, but densities whose modes do not coincide with the target would cause doubts about whether the attempts have been effective or whether another target better explains the data. We call this *normal-like-or-better* (NLOB) performance effectiveness. It is next necessary to obtain the analog of this criterion for the efficiency performance data, Y_{rj}, relevant to the present context.

If $\mathbf{x}$ is distributed as $N(\boldsymbol{\mu}, \boldsymbol{\Sigma})$ on $\Re^n$ then it is well known that the quadratic form, $w(\mathbf{x}) = (\mathbf{x} - \boldsymbol{\mu})' \boldsymbol{\Sigma}^{-1} (\mathbf{x} - \boldsymbol{\mu})$ has the gamma distribution $g(\alpha, \beta)$, where $\alpha = n/2$ and $\beta = 2$. This distribution is also called the Chi-square distribution with n degrees of freedom (see Law and Kelton, 1982). We may note that for this case, $w(\mathbf{x})$ is in the nature of a squared-distance from the target, here the singleton set $\{\boldsymbol{\mu}\}$. It is useful to derive this result by way of VDR. Let $w(\mathbf{x})$ be a continuous convex function on $\Re^n$ with range $[0, \infty)$; and let $g(w)$ be a pdf on $[0, \infty)$. Suppose that for each value of $u \geq 0, \mathbf{x}$ is uniformly distributed on the set $\{x : w(\mathbf{x}) = u\}$. Consider the process of sampling a value of u according to the $g(w)$ density and then sampling a vector, $\mathbf{x}$, according to the uniform distribution on the set $\{\mathbf{x} : w(\mathbf{x}) = u\}$. Next let $f(\mathbf{x})$ be the pdf of the resulting $\mathbf{x}$ variates on $\Re^n$. Finally let $A(u)$ be the Lebesgue measure of the set $\{\mathbf{x} : w(\mathbf{x}) \leq u\}$. Assuming that $A(u)$ is differentiable on $[0, \infty)$ with $A'(u)$ strictly positive, then Theorem 1.2 shows that

$$f(\mathbf{x}) = \phi(w(\mathbf{x})) \quad \text{where} \quad g(w) = \phi(w)/A'(w). \tag{8.12}$$

We may use this result to derive a very general density class for performance related to squared-distance type error measures. The set $\{\mathbf{x} : (\mathbf{x} - \boldsymbol{\mu})'\boldsymbol{\Sigma}^{-1}(\mathbf{x} - \boldsymbol{\mu}) \leq u\}$ has volume, $A(u)$, given by $A(u) = \alpha_n|\boldsymbol{\Sigma}|^{1/2}u^{\frac{n}{2}}$ where $\alpha_n = \frac{\pi^{\frac{n}{2}}}{\frac{n}{2}\Gamma(\frac{n}{2})}$ (Fleming, 1977), so that $A'(u) = \frac{n}{2}\alpha_n|\boldsymbol{\Sigma}|^{1/2}u^{\frac{n}{2}-1}$. The gamma $g(\alpha, \beta)$ density is given by

$$g(u) = (\Gamma(\alpha)\beta^\alpha)^{-1}u^{\alpha-1}\exp\{-\frac{u}{\beta}\}. \tag{8.13}$$

Therefore (8.12) implies that if $w(\mathbf{x}) = (\mathbf{x} - \boldsymbol{\mu})'\boldsymbol{\Sigma}^{-1}(\mathbf{x} - \boldsymbol{\mu})$ and $g(u) = g(\alpha, \beta)$, then the corresponding $f(\mathbf{x})$, which we now rename as $\Psi(\mathbf{x}) = \Psi(\mathbf{x}; n, \alpha, \beta)$, is given by

$$\begin{aligned}
\Psi(\mathbf{x}) &= \Gamma(\frac{n}{2})(\pi^{\frac{n}{2}}\Gamma(\alpha)\beta^\alpha)^{-1}\left[(\mathbf{x} - \boldsymbol{\mu})'\boldsymbol{\Sigma}^{-1}(\mathbf{x} - \boldsymbol{\mu})\right]^{\alpha-\frac{n}{2}} \\
&\quad \exp\{-\frac{1}{\beta}(\mathbf{x} - \boldsymbol{\mu})'\boldsymbol{\Sigma}^{-1}(\mathbf{x} - \boldsymbol{\mu})\}
\end{aligned} \tag{8.14}$$

For this density class we have the following observations:

(i) If $\alpha = \frac{n}{2}$ and $\beta = 2$ then $\Psi(\mathbf{x})$ is the multivariate normal density, $N(\boldsymbol{\mu}, \boldsymbol{\Sigma})$.

(ii) If $\alpha = \frac{n}{2}$ and $\beta \neq 2$ then $\Psi(\mathbf{x})$ is steeper or flatter than $N(\boldsymbol{\mu}, \boldsymbol{\Sigma})$ according to whether $\beta < 2$ or $\beta > 2$, respectively. We may call these densities the *normal-like* densities.

(iii) If $\alpha < n/2$ then $\Psi(\mathbf{x})$ is unbounded at its mode but may be more or less steep according to the value of β. We call this class the *better-than-normal-like* density class.

(iv) If $\alpha > n/2$ then $\Psi(\mathbf{x})$ has zero density at the target $\boldsymbol{\mu}$, and low values throughout neighborhoods of $\boldsymbol{\mu}$. This suggests that attempts at the target are not effective. The data may have arisen in pursuit of a different target or simply not be effective for any target.

For densities in category (iii), the unbounded mode concentrates more probability near the target and suggests a higher level of expertise than that evidenced by the finite-at-mode $N(\boldsymbol{\mu}, \boldsymbol{\Sigma})$ class. It seems reasonable to refer to α in this context as the *expertise, mode,* or *target effectiveness* parameter; while β is a scale or precision parameter. Thus, if $\alpha \leq n/2$ we call $\Psi(\mathbf{x})$ the normal-like-or-better performance density. To summarize, if attempts at a target set in $\Re^n$ have a basic squared-distance error measure and this measure is distributed according to the $g(\alpha, \beta)$ density with $\alpha \leq n/2$

then the performance with respect to this target set is *normal-like-or-better* (NLOB).

For the present estimation model, the target set is $\{Y \in \Re^4 : \sum_{r=1}^{R} a_r Y_r = 1, Y_r \geq 0$ for all $r\}$ rather than the point set $\{\boldsymbol{\mu}\}$. If $\sum_{r=1}^{R} a_r Y_r = v_j$ then the distance of Y_{rj} from the target set is $(1 - v_j) \| a \|^{-1}$. Since $0 \leq v_j \leq 1$ we employ the transformation $w = (- \ln v)^2 = (\ln v)^2$. This transformation has the properties that $w \cong (1 - v)^2$ near $v = 1$ and $w \in [0, \infty)$. Therefore $\frac{w}{\|a\|^2} = \frac{(\ln v)^2}{\|a\|^2}$ is an approximate squared-distance measure near the target set. Since the $\| a \|^2$ term is a scale factor, it can be absorbed into the β parameter of $g(\alpha, \beta)$. We therefore consider the NLOB effectiveness criterion to hold if w has the $g(\alpha, \beta)$ density with $\alpha \leq 2$. That is, such performance is analogous to that of unbiased normal-like-or-better distributed attempts at a target in $\Re^n$.

There is one additional consideration before applying this effectiveness criterion to the present data. In the LP estimation model MPE, at least one efficiency, v_j, must be unity (and hence $w_j = 0$). This is because at least one efficiency inequality constraint must be active in an optimal solution of the MPE model. We therefore consider the model for the w_j to be

$$p\delta(0) + (1 - p)g(\alpha, \beta) \tag{8.15}$$

where p is the frequency of zero values beyond one (here $p = 3/62 = 0.048$ from Table 8.1), and $\delta(0)$ is the degenerate density concentrated at $w = 0$. For this data, we regard the NLOB criterion to hold if it holds for the gamma density after omitting the zeroes. Thus, when the $g(\alpha, \beta)$ density is fitted to the strictly positive w-values then NLOB requires that $\alpha \leq 2$. For the data of $w_j = (\ln v_j)^2$ based on Table 8.1, column 5, the parameter value estimates obtained by the Method of Moments (see, for example, Bickell and Doksum, 1977) are $\alpha = 1.07$ and $\beta = 0.32$. This method was chosen because the BestFit^{TM} (1995) software experienced difficulty in convergence using its default maximum likelihood estimation procedure. The method of moments estimates parameters by setting theoretical moments equal to sample moments. For the gamma density, $\mu = \alpha\beta$, and $\sigma^2 = \alpha\beta^2$. If $\bar{w}$ and s^2 are the sample mean and variance of the positive w_j values, then the α and β estimates are given by

$$\hat{\alpha} = \frac{\bar{w}^2}{s^2} \quad \text{and} \quad \hat{\beta} = \frac{s^2}{\bar{w}} \tag{8.16}$$

Tests of fit of the w_j data to the $g(\alpha = 1.07, \beta = 0.32)$ density were carried out using BestFit^{TM} (1995). All three tests provided there, the Chi-square, Kolmogorov-Smirnov, and the Anderson-Darling indicated acceptance of the gamma model with confidence levels greater than 0.95. In addition, for each of these tests, the gamma model was judged best fitting (rank one) among the densities in the library of BestFit^{TM}. We therefore conclude that the NLOB criterion was met. The NLOB criterion is important in establishing whether the estimated cost model is a plausible goal of the units being studied. The MPE model will produce estimates for any arbitrary set of Y_{rj}-data. However, if the resulting v_j were, for example, uniformly distributed on $[0, 1]$, there would be little confidence in the estimated model.

The gamma assumption for $g(v)$ is not critical. This assures, together with an acceptable alpha parameter value, that the performance vectors have a normal-like distribution with mode equal to the target set estimated by the model. A more general requirement with the same effect is that the performance vectors be distributed as unimodal with mode coinciding with the target set. If, for example, $g(v)$ has a Weibull density, say, then the solution of (8.12) might be reversed to obtain the $f(\mathbf{x})$-density. The resulting density can be checked for unimodality and its mode can, in principle, be directly compared to the target set. However, in these situations it should be more convenient to use the general TMA criterion of Chapter 7.

8.5 Discussion

In this section, we briefly discuss the weights flexibility issue in DEA, longitudinal use of the approach, and implications for the single driver - single cost pool case. We also mention some limitations and questions needing further research. For a discussion of the managerial significance of the results, the reader is referred to Troutt *et al.* (2000) or Troutt *et al.* (2003b).

8.5.1 *Weights Flexibility*

From Table 8.1, it is clear that the proposed method is a more stringent measure of efficiency than DEA. Comparing the proposed method only to DEA, it can be seen that for all units, efficiency scores are largest for DEA and smallest for the proposed method. In addition, in each case except Bradford, the DT-efficiencies were between these bounds. This is as expected since DEA permits maximal weights flexibility and the proposed method permits no weights flexibility. Thus when used with DEA, a range estimate of the efficiency of each unit is obtained. The proposed method assumes that all units are comparable and therefore should have the same minimal unit cost goals.

8.5.2 *Longitudinal Data*

Suppose the data Y_{rt} are given over time periods indexed by t for a single business unit. Then the MPE model with index j replaced by t might be applied as a model for internal benchmarking. First, it would be necessary to adjust all the x_t-cost pool figures, and resulting Y_{rt}-data to reflect current dollars using a cost index. This assumes that the estimated a_r cost rates are in terms of current dollars. Then the estimated a_r may be interpreted to be the costs achieved by the unit during its most efficient observation period or periods. The resulting v_t suggest periods of more or less efficiency, and would be a useful source for self-study aimed at productivity and process improvements. The comparability issue for the units under comparison should be easier to accept in this case. However, process or technology and market or other environmental changes during the data time span could be problematical. A more complete discussion of limitations for this case is left for specific future application case studies.

8.5.3 *The Single Driver and Single Cost Pool Case*

The MPE model for this case simplifies to $\max \sum aY_j$, s. t. $aY_j \leq 1$, for all j, and $a \geq 0$. The solution of this model is clearly $a^o = \min Y_j^{-1}$. It may be verified that the NLOB criterion requires $\alpha^o \leq \frac{1}{2}$ in this case. If this condition fails to hold then this minimum value may be unreasonably low, perhaps due to an outlier. Deletion of one or a few tentative outliers would be well supported if the remaining data pass the NLOB test. Otherwise no credible a^o-estimate is forthcoming from the present method.

8.5.4 *Limitations and Further Research*

In order to more fully parallel existing OLS theory for model aptness testing, attention should be given to potential outliers, independence of the

v_j, transformations and constancy of the distribution of the v_j from trial to trial (analogous to homoscedasticity in OLS theory); see, for example, Madansky (1988) and Neter *et al.* (1985). Theory developments for these issues are not yet available for the MPE approach and would be worthwhile for future research. TDF also discuss what we have called comparability of these units. A concern was noted relative to activity four whose monetary driver level might have been affected by the prosperity of the community being served. That is, units with above average property values might be considered as being unfairly compared to the others. Other things being equal, units with an inappropriately inflated value of a driver level would be expected to exert a downward influence on the corresponding estimate in model MPE. We believe this kind of potential incomparability might be avoided by use of a property value index for future research.

We have assumed that the benchmark unit costs should be strictly positive in this chapter. As noted above it, it may be possible that for some situations and units that an activity can be achieved as a cost-free by-product of others. This might not necessarily be inconsistent with the comparability assumption when the practice or process responsible for such an advantage is transferable to the other units. The unmodified MPE model appears capable of indicating such situations. However, it should be validated on a known case of this type for further research.

8.6 Some Final Remarks on the Single Cost Pool Case

So far, we have proposed a new method for estimating cost efficiencies and benchmark unit costs in the case of a single cost pool. The results provide a new tool for certain benchmarking studies in activity-based costing. The estimated costs provide plausible operational goals for the management of the units being compared. This method also provides efficiency measures and suggests which organizational units or periods are more or less efficient, as well as an estimate of the degree of such inefficiency. Efficient units or periods provide benchmarks for imitation by other units or can be studied for continuous improvement possibilities. The resulting estimates were compared with the average costs obtained by another previous method. The estimated benchmark cost rates were uniformly and strictly lower than the average rates, consistent with their definitions and providing a good measure of face validity.

8.7   Benchmark Cost Matrices

Benchmark costs may be defined as the apparent lowest attainable costs based on observed performances of a set of comparable units over a number of periods. This concept is operationalized using what we call technically efficient weak and strong benchmark cost matrices. From the set of such matrices, a specific estimate is obtained as one that can be attributed to the most efficient unit(s) and period(s) in the data. The multiple costs and drivers case has not been addressed frequently. In a study on airline industry data, Banker and Johnston (1993) used multiple regression models for similar data to estimate what might be considered average rather than best practice costs. However, no previous consideration of the benchmark costs focus in this context is known to the authors.

In the first part of this chapter, we considered the case of a multiple driver-single cost pool for data in which driver levels were nondiscretionary. In the rest of this chapter, the focus is on discretionary cost drivers and multiple cost pools. It is not known whether the earlier method can be modified for the present assumptions and the method proposed next is independent of the earlier approach. We now introduce some basic notations and definitions.

Indexes:

$$i \;=\; 1,\ldots,K - \text{identifies the cost pool or type.}$$

$$r \;=\; 1,\ldots,R - \text{identifies the cost driver or output type.}$$

$$j \;=\; 1,\ldots,J - \text{identifies the department, more generally, the operational unit.}$$

$$t \;=\; 1,\ldots,T - \text{identifies the period.}$$

Data:

X_i^{jt} —   cost pool amounts. The indexed column vector form of this array is denoted $\mathbf{X}^{jt}$.

Y_r^{jt} —   driver or output amounts. $\mathbf{Y}^{jt}$ is the indexed column vector form.

Table 8.2: Example of a Multiple Cost Pool and Multiple Cost Driver Data Table

		Driver Levels - Y				Costs - X	
		Credit hrs.				Dollars	
	Dept.	Gen'l Studies	BS	MS	PhD	Dept. Bugdet	College Overhead
No.	Name	$r = 1$	$r = 2$	$r = 3$	$r = 4$	$i = 1$	$i = 2$
1	ACCT	56	2781	571	203	1706605	185797
2	ADMS	28	5978	946	568	2406355	429622
3	ECON	2889	408	562	159	1236807	214367
4	FIN	0	2758	361	329	1365231	184665
5	MKTG	25	3019	134	169	1354080	189177

Table 8.2 shows example data of the type motivating this study and identified by these notations. This is the data for one ($T = 1$) annual period for a business college consisting of five ($J = 5$) academic departments: Accounting, Administrative Sciences, Economics, Finance and Marketing. Driver levels, measured as credit hours of instruction, are shown for each of four ($R = 4$) academic instructional activities: General Business Studies, Bachelor of Science, Master of Science and the PhD program. For each department, costs are allocated to two ($K = 2$) cost pools: Department Budget and College Overhead, respectively. Table 8.2 is an excerpt from a larger data set prepared for the years 1987-1995 by the Office of Institutional Research and Decision Support at the first author's university. Longitudinal cost data of this kind would need adjusted by an inflation index. The Higher Education Price Index (HEPI) is one such index available from the website of Research Associates of Washington.

Notational conventions: Matrix transpose is denoted by the prime symbol. We use **e** to denote the all-units column vector of appropriate dimension for context of usage; **0** is the zero vector and **I** is the identity matrix.

Parameters to be estimated:

A $=$ $\{a_{ir}\}$ – the $K \times R$ matrix of benchmark costs discussed in detail below. The (i, r)-th element is the type-i cost in dollars per one credit hour of instructional activity, r.

$$\boldsymbol{\pi} \;=\; \{\pi_r\} - \text{a column vector of normalized relative values of outputs.}$$

The π_r-values may also be considered as relative priorities for the different instructional activities. No other use of $\boldsymbol{\pi}$ is made in this chapter. It may be regarded as merely facilitating definition of the auxiliary target model. We later discuss the more general case in which departments may have differing objective coefficient vectors, $\boldsymbol{\pi}^j$.

Variables:

$$\mathbf{y}^{jt} \;=\; y_r^{jt} - \text{level of driver } r \text{ in credit}$$
$$\text{hours for unit } j \text{ in period } t.$$

8.7.1   *Technically Efficient Benchmark Cost Matrices*

We call matrix $\mathbf{A}$, a benchmark cost matrix if $\mathbf{A}\mathbf{Y}^{jt} \leq \mathbf{X}^{jt}$ for all j and t and the elements of $\mathbf{A}$ are nonnegative. The motivation for the nontrivial condition is as follows. If each department over all periods had generated costs at the rates given by a benchmark cost matrix then its cost pool values could not have been larger than the observed values.

Definition 8.1: Let $\mathbf{A} = \{a_{ir}\}$ be a benchmark cost matrix. If for each i, $\sum_{r=1}^{R} a_{ir} Y_r^{jt} = X_i^{jt}$ holds for some (j,t), then we call $\mathbf{A}$ a *technically efficient benchmark cost matrix*.

This definition limits the admissible class of estimates to those benchmark cost matrices, which fit the data as tightly as possible. In general, the problem of finding a linear programming system that fits a set of data may be called *linear programming system identification*. Some other work in that area is given in Troutt *et al.* (2003c).

8.7.2   *The Target Linear Programming Model*

We assume that there exists a vector $\boldsymbol{\pi}$, such that π_r is the relative value per credit hour of type r, and that these relative values are constant across all departments in the college, as well as constant over the periods of data considered. Suppose that $\mathbf{A}$ and $\boldsymbol{\pi}$ are known. Then, having been given a budget of X_i^{jt}, each department, j, could have determined optimal driver levels in period, t, according to the following respective *target primal linear programming models,*

TPLP(j,t):

$$\max \quad \boldsymbol{\pi}' \mathbf{y}^{jt} \tag{8.17}$$

$$s.t. \quad \mathbf{A}\mathbf{y}^{jt} \leq \mathbf{X}^{jt} \text{ for all } j \text{ and } t, \text{componentwise} \tag{8.18}$$

$$y_r^{jt} \geq 0 \text{ for all } r, \ j \text{ and } \ t. \tag{8.19}$$

These problems are $J \times T$ in number. Their solutions provide best practice or ideal driver level targets for department j in period t in terms of the unknown benchmark costs and output or driver relative valuations, $\boldsymbol{\pi}$. It will be shown that observed performances can be compared to these ideals. We first assume for simplicity that the driver levels are fully discretionary. Academic departments control credit hours by choosing section sizes and numbers, course entry and passing requirements, recruiting activities and staffing decisions, among others. Later we consider a weaker assumption, which limits the variability of driver levels by reference to historical patterns in the data. This requires that an additional set of constraints be calculated from the data and added to the above models. A further modification in which cost pools are also discretionary will be discussed below. It is not necessary to assume that the departments did in fact, use such models explicitly. Moreover, use of such a model would not be possible without actual knowledge of the unknown costs. Rather, we assume that the departments have subjectively and approximately solved this or a strategically equivalent model in their past actions.

In the next section, we derive the MPE model for the estimation of $\boldsymbol{\pi}$ given the matrix $\mathbf{A}$. The overall strategy for this estimation can now be outlined. We construct a representation of the space of benchmark $\mathbf{A}$-matrices, which is suitable for genetic search. Density models and a likelihood score for the observed data are developed for each choice of the $\mathbf{A}$-matrix and associated $\boldsymbol{\pi}$-given-$\mathbf{A}$ estimate. Then the likelihood score may be used as a fitness function for genetic search over the representation space. Thus, we seek jointly maximum likelihood estimates of $\boldsymbol{\pi}$ and $\mathbf{A}$.

8.8 MPE Estimation of the Objective Coefficient Vector

To employ the MPE technique, performance efficiency scores for each j and t must be derived in terms of the unknown parameters. For construction of the performance efficiencies, assume that a candidate estimate, $\mathbf{A}$, is given and let $\mathbf{y}^{jt*}$ be any optimal solution vectors for the models, $\mathbf{TPLP}$(j,t). Then $\boldsymbol{\pi}'\mathbf{Y}^{jt}$ is the value of the observed driver levels and $\boldsymbol{\pi}'\mathbf{y}^{jt*}$ is the model-optimal value. Therefore, the scalar values

$$\eta^{jt} = \boldsymbol{\pi}'\mathbf{y}^{jt}/\boldsymbol{\pi}'\mathbf{y}^{jt*} \tag{8.20}$$

yield a set of performance efficiency ratios suitable for use with the MPE approach. For simplicity, the dependence on $\mathbf{A}$ and the data is not indicated in (8.20). The right hand side can be simplified by considering the corresponding dual problems, $\mathbf{DPLP}$(j,t). Let the column vectors, $\boldsymbol{\xi}^{jt}$, be the dual variables associated with (8.20) . These dual problems, $J \times T$ in number, are given by

$$\mathbf{DPLP}(\text{j,t}): \quad \min \quad \mathbf{X}^{jt'}\boldsymbol{\xi}^{jt} \text{ s.t. } \mathbf{A}'\boldsymbol{\xi}^{jt} \geq \boldsymbol{\pi}, \ \xi_i^{jt} \geq 0 \text{ for all } i,j \text{ and } t.$$

Let $\boldsymbol{\xi}^{jt*}$ be any collection of optimal solutions. Then by linear programming duality we have $\boldsymbol{\pi}'\mathbf{y}^{jt*} = \mathbf{X}^{jt}\boldsymbol{\xi}^{jt*}$ for all j and t. Therefore (8.20) may be written as

$$\eta^{jt} = \boldsymbol{\pi}'\mathbf{y}^{jt}/\mathbf{X}^{jt'}\xi^{jt*} \tag{8.21}$$

It may be noted that $\boldsymbol{\pi}$ can be scaled arbitrarily in these fractions since a proportional increase in $\boldsymbol{\pi}$ is matched by the same proportional increase in the $\boldsymbol{\xi}^{jt*}$. This has been called the *radial change property* of linear programming and is discussed further in Troutt *et al.* (2000). We therefore choose the scaling, $\| \boldsymbol{\pi} \|^2 = 1$, in order to select one possible solution. However, due to a maximization objective and positive data values, this can be relaxed to the constraint $\| \boldsymbol{\pi} \|^2 \leq 1$ along with the nonnegativity constraints $\pi_r \geq 0$ in the estimation sub-problem below.

Next, it is clearly necessary that $\boldsymbol{\pi}'\mathbf{Y}^{jt} \leq \boldsymbol{\pi}'\mathbf{y}^{jt*} = \mathbf{X}^{jt}\boldsymbol{\xi}^{jt*}$ for all j and t, which can be regarded as requiring that $\eta^{jt} \leq 1$ for all j and t.

A modification of the dual constraints, $\mathbf{A}'\boldsymbol{\xi}^{jt} \geq \boldsymbol{\pi}$, to $\mathbf{A}'\boldsymbol{\xi}^{jt} = \boldsymbol{\pi}$ can be argued as follows. All data values Y_r^{jt} are positive for this study. It is reasonable to suppose therefore and from knowledge of the real system that the corresponding optimal but unobserved y_r^{jt*} should be positive also. By choosing the equality form of the dual constraints, such positive solutions are therefore expected to be promoted in cases of alternative optima and degenerate solutions.

Collecting the above considerations yields the following mathematical programming model for the estimation of π given $\mathbf{A}$.

MPE$(\pi|\mathbf{A})$:

$$\max \quad \sum_j \sum_t \pi'\mathbf{Y}^{jt}/\mathbf{X}^{jt\prime}\boldsymbol{\xi}^{jt} \tag{8.22}$$

$$\text{s.t.} \quad \mathbf{A}'\boldsymbol{\xi}^{jt} = \pi \text{ ,for all } j \text{ and } t \tag{8.23}$$

$$\pi\mathbf{Y}^{jt} \leq \mathbf{X}^{jt}\boldsymbol{\xi}^{jt} \text{ ,for all } j \text{ and } t \tag{8.24}$$

$$\| \pi \|^2 \leq 1 \tag{8.25}$$

$$\pi_r, \ \xi_i^{jt} \geq 0, \text{ for all } r, i, j \text{ and } t \tag{8.26}$$

Except for the norm constraint $\| \pi \|^2 \leq 1$, model **MPE$(\pi|\mathbf{A})$** is a programming problem having a sum of linear fractions objective function with linear constraints and would therefore belong to a class of problems for which algorithms have been proposed. For example, Falk and Palocsay (1992) proposed an algorithm in which constraints on various feasible sets are iteratively tightened. Quesada and Grossman (1995) have proposed a branch and bound algorithm. However, the difficulty of modifying such algorithms to accommodate the norm constraint suggested the use of a general nonlinear solver for convex constraint sets. We therefore suggest a generalized reduced gradient algorithm code.

8.9   A Likelihood Model for the Data

One potential strategy for completing the estimation problem would be to adjoin the benchmark matrix definition conditions, $\mathbf{A}\mathbf{Y}^{jt} \leq \mathbf{X}^{jt}$,componentwise, to the **MPE$(\pi|\mathbf{A})$** model and consider its optimization to extend over the elements of $\mathbf{A}$ as well as those of π. However, constraints (8.23) then become nonconvex quadratic ones and make that approach problematical. In addition, it is easily seen that such a problem may have an essentially trivial solution. Namely, the $\mathbf{A}$-matrix all of whose rows are scalar multiples of π would suffice. Hence, it does not seem obvious how to require that possibly, if not likely, distinct rows of the $\mathbf{A}$-matrix are required. These observations may also help to explain why the MPE technique used alone does not suffice for estimating both π and $\mathbf{A}$-parameter arrays. The MPE approach assumes that the $\mathbf{Y}^{jt}$ were chosen to optimize efficiency. This is only one aspect in the present setting. Namely, given a certain level of efficiency attainment, a cost driver vector only needs to be chosen randomly in the space of possible values having the same performance efficiency. Thus, the present process is partly efficient performance oriented and partly random. Therefore, a hybrid estimation strategy is developed by constructing a pdf on the set of feasible data vectors, from which a likelihood score can be assigned to each candidate $\mathbf{A}$-matrix and associated $\boldsymbol{\pi} = \boldsymbol{\pi}(\mathbf{A})$-estimate. By solution of model $\mathbf{MPE}(\boldsymbol{\pi}|\mathbf{A})$, the efficiency ratios $\eta^{jt} = \boldsymbol{\pi}'\mathbf{Y}^{jt}/\mathbf{X}^{jt}\boldsymbol{\xi}^{jt*}$ may then be computed. The η^{jt} are distributed over the interval $[0,1]$, and a specific pdf $g(\eta)$ can be fitted to them. This pdf may be used to construct a pdf on the set of driver vectors as developed next.

8.9.1   *Density and Likelihood Models*

The following further definitions are needed. Let F^{jt} be the sets of feasible driver vectors, $\mathbf{y}$, for problems $\mathbf{TPLP}$(j,t), respectively and let z^{jt*} be the optimal objective function value for those same problems. Next, let $E^{jt}(\eta) = \{\mathbf{y} \in F^{jt} : \boldsymbol{\pi}\mathbf{y} = \eta z^{jt*}\}$. These sets contain those feasible driver vectors that have performance efficiency score η. Note that these sets also depend on the data, $\mathbf{X}^{jt}$, through problem, $\mathbf{TPLP}$(j,t), but this is left out of the notation for simplicity. Next, define $W^{jt}(\eta) = \cup_{\eta' \geq \eta} E^{jt}(\eta')$ as the sets of driver vectors that are feasible for problem $\mathbf{TPLT}$(j,t) and which also have efficiency scores of η or greater. Now define the functions $V^{jt}(\mathbf{y}) = \boldsymbol{\pi}\mathbf{y}/z^{jt*}$. These are defined on the set of feasible driver vectors for problems, $\mathbf{TPLP}$(j,t), and give the performance efficiency values of such feasible $\mathbf{y}$ vectors for the instances defined by j and t. Finally, let $g_\eta(\eta)$ be the pdf of the population from which the η^{jt} form a sample. As in Troutt *et al.* (2000) and in Chapter 7 above, we model this pdf as follows. Let $\omega = (\ln \eta)^2$. It is assumed that ω has a gamma pdf

$$g_\omega(\omega) = \{\Gamma(\alpha)\beta^\alpha\}^{-1}\omega^{\alpha-1}\exp\{-\omega/\beta\}. \tag{8.27}$$

with shape parameter α and scale parameter β . Following calculation of the η^{jt} and transformation to ω^{jt}, these parameters can be estimated using the method of moments. Namely, if $\bar{\omega}$ and s^2 are the sample mean and variance, respectively, of the positive values, then the α and β estimates were taken as given by $\alpha^o = (\bar{\omega})^2/s^2$ and $\beta^o = s^2/\omega^o$. For the pdf model $f^{jt}(\mathbf{y})$ we may use the technique of section 6.6 above. In the present context, we have

$g(\eta), 0 \leq \eta \leq 1$, as the pdf for the η^{jt} efficiency scores. Also, let z^{jt*} be the optimal objective function values for problems, **TPLT**(j,t). Then it follows that the densities, $f^{jt}(\mathbf{Y})$ are given by

$$f^{jt}(\mathbf{Y}^{jt}) = \| \, \boldsymbol{\pi} \, \| g(V^{jt}(\mathbf{Y}^{jt}))[z^{jt*}L(E^{jt}(\eta^{jt}))]^{-1}. \qquad (8.28)$$

Finally, we obtain the likelihood of the whole data sample, assuming independence, as

$$\prod_{j=1,\ldots,J} \prod_{t=1,\ldots,T} f^{jt}(\mathbf{Y}^{jt}). \qquad (8.29)$$

In case $L(E^{jt}(\eta)) = 0$ for some η^{jt}, the corresponding $f^{jt}(\mathbf{Y}^{jt})$ may be undefined as shown by (8.28). We then propose the following heuristic, which we call *indented likelihood*. If any $f^{jt}(\mathbf{Y}^{jt})$ is undefined, its value is replaced by the largest of those that are defined. Then (8.29) is calculated by the revised values. This heuristic has the following properties. A sample cannot have undefined likelihood unless for every j and t, $\mathbf{Y}^{jt} = \mathbf{y}^{jt*}$ holds. Except for that case, the likelihood will be defined and may have any value in $[0, \infty)$. Such cases can only occur if the associated $\eta = 1.0$. An alternative approach is based on the premise that attainment of full (unity) efficiency is always unreasonable in reality and that efficiency score sets should be adjusted downward in some appropriate fashion. Two *simulation-optimization* approaches to efficiency score adjustments of this kind are given in Troutt *et al.* (2001). The simulation-optimization technique performs a genetic algorithm search over possible trial values of the maximum efficiency. For each trial value, simulation is used to estimate the maximum efficiency for a sample of fixed size. The goal of the search is to find a trial value that matches the mean of such simulated maximum efficiency values.

The next section discusses representation of the space of technically efficient benchmark cost matrices for genetic search and some related technical issues.

8.10 Genetic Search Space and Technical Issues

8.10.1 *Representation of the A-Matrices*

Let $\mathbf{\Lambda} = \lambda_{ir}$ be a matrix for which $\lambda_{ir} \geq 0$ for all r and i with $\sum_{r=1}^{R} \lambda_{ir} = 1$ for each i. We call $\mathbf{\Lambda}$ the constraint generator matrix. Let κ_i be scalars such that a_{ir}, the elements of $\mathbf{A}$, are given by $a_{ir} = \kappa_i \lambda_{ir}$. In order for

$\mathbf{AY}^{jt} \leq \mathbf{X}^{jt}$ to hold for all j and t, it is necessary that $\sum_{r=1}^{R} \kappa_i \lambda_{ir} Y_r^{jt} \leq X_i^{jt}$ for all j and t. Furthermore, we desire the largest such κ_i-values. Otherwise, strict inequality would hold for all i, j, and t, so that no candidate $\mathbf{A}$-matrix could be technically efficient. These considerations yield

$$\kappa_i = \min_{j,t}\{X_i^{jt} / \sum_{r=1}^{R} \lambda_{ri} Y_r^{jt}\}. \tag{8.30}$$

For representation of the $\mathbf{\Lambda}$-matrix, it is further convenient to define variables, $u_{ir}, r = 1, \ldots, R - 1$, as follows. First, we require that $0 \leq u_{ir} \leq 1$ and $u_{iR} = 1$ for all i and r. Then

$$
\begin{aligned}
\lambda_{i1} &= u_{i1}, \\
\lambda_{i2} &= (1 - \lambda_{i1})u_{i2}, \\
\lambda_{i3} &= (1 - \lambda_{i1} - \lambda_{i2})u_{i3}, \\
&\quad \cdots \\
\lambda_{ir} &= (1 - \sum_{n=1}^{r-1} \lambda_{in})u_{ir}.
\end{aligned}
\tag{8.31}
$$

This enables the genetic search to be performed over the unit-hypercube defined by the u_{ir} ranges.

To summarize, genetic search can be performed on the space of the variables u_{ir} with $i = 1, 2$ and $r = 1, 2, 3$. Then these values may be converted to the $\mathbf{\Lambda}$-matrix values. Next, the κ-values are computed from the data and the corresponding candidate $\mathbf{A}$-matrix instance is formed. Problem $\mathbf{MPE}(\pi|\mathbf{A})$ is then solved and the likelihood is computed as a fitness score for genetic search over the matrices $\mathbf{U} = \{u_{ir}\}$.

8.10.2 *Extreme Points and Set Volumes*

For each instance of the $\mathbf{A}$-matrix evaluated in the genetic search, it is necessary to evaluate the volumes of the sets $L(E^{jt}(\eta))$ and associated extreme points. The sets, $E^{jt}(\eta)$, which depend on π and $\mathbf{A}$, are polyhedral convex sets. Methods for calculating volumes of such sets are discussed in Fleming (1977), Cohen *et al.* (1979), Berger (1987) and Verschelde *et al.* (1994). These methods require enumeration of the extreme points for which algorithms have been suggested by Walker (1973), Schactman (1974) and Deusing (1977). A survey is given by Mattheis and Rubin (1980). Recent surveys and practical guides for computing volumes of convex polytopes have been given by Lawrence (1991) and Büeler *et al.* (1998).

8.10.3 *Genetic Search for All Parameters*

Two potential variations of our estimation procedure can be considered. First, the search space might be enlarged to include generator vectors similar to the λ-vectors for representation of the π-vector. With this approach, problem $\mathbf{MPE}(\pi|\mathbf{A})$ would be simplified as π would be specified for each population element in the search. That is, it would not be necessary to solve the $\mathbf{MPE}(\pi|\mathbf{A})$ problem at each iteration. However, the $\boldsymbol{\xi}^{jt*}$ would still need to be computed by the simplified $\mathbf{MPE}(\pi|\mathbf{A})$ problem. The dimension of the search space would be enlarged without significant savings in the programming step and this approach cannot be recommended. Similarly, the search space might be further enlarged to include representation of the $\boldsymbol{\xi}^{jt*}$ vectors. Since this would greatly increase the dimension of the search space and would also require enforcement of the constraints in problem $\mathbf{MPE}(\pi|\mathbf{A})$, neither can it be recommended.

8.11   Validation Issues

8.11.1 *Use of the Target-Mode Agreement Criterion*

The TMA criterion of Section 7.4.1 can also be applied in this case. The target set for the (j,t)-th trial in the present context is the set of optimal solutions for problems $\mathbf{TPLP}(j,t)$ identified by $\eta = 1.0$. These target sets may be single points in case of unique optimal solutions or convex polyhedral sets more generally. We define the distance between the data point $\mathbf{Y}^{jt}$ and the target set to be the distance between this data point and the hyper-plane, $\pi'\mathbf{y} = \pi'\mathbf{y}^{jt*} = z^{jt*}$. For the data point, $\mathbf{Y}^{jt}$, we have $\pi'\mathbf{Y}^{jt*} = \eta^{jt}z^{jt*}$. This distance is given by $(1 - \eta^{jt})z^{jt*}\| \pi \|^{-1}$. However, we employ the transformation, $(-\ln\eta)^2 = (\ln\eta)^2 \cong (1-\eta)^2$ near $\eta = 1$, in order that distances be defined in the interval $[0,\infty)$. Thus, we obtain the final squared distance measure as

$$\omega^{jt} = \{(\ln\eta^{jt})z^{jt*}\| \pi \|^{-1}\}^2 \tag{8.32}$$

8.11.2 *The Comparison to Regression Coefficients Criterion*

Considering $\mathbf{A}^{jt}\mathbf{Y}^{jt} = \mathbf{X}^{jt}$ for all j and t suggests that regression through the origin row by row should lead to estimation of a kind of average, $\mathbf{A}^+$, of the $\mathbf{A}^{jt}$. In the case of a single cost pool, matrix A has just one row. Hence, this is the matrix analog of the suggestion by Dyson and Thanassoulis (1988) discussed earlier in Section 8.3. If $\mathbf{A}^o$ is an estimated benchmark cost matrix, then one should expect that $\mathbf{A}^o \leq \mathbf{A}^+$ componentwise. Hence, if this test holds we consider that the estimate $\mathbf{A}^o$ exhibits good face validity as a benchmark cost estimate.

8.12 Miscellaneous Issues

8.12.1 *Data Requirements Issues*

A potential limitation is the number of parameters to be estimated in comparison to the number of data observations available. For the example of Table 8.2, there are 12 parameters to be estimated in total as elements of π and $\mathbf{A}$. The 10k Rule from regression methodology (See Harrell 2001 and Kleinbaum *et al.*, 1997) suggests that there should be at least ten times as many data points as there are parameters to be estimated. Borrowing this heuristic from regression yields the requirement of 120 observations. However, it can be argued that the 10k Rule is likely to be too severe for models of this kind. Namely, all parameters are nonnegative in the model. In addition, one degree of freedom, i.e. one dimension of π may be deducted due to the norm constraint. Similarly, the matrices are generated from the Λ-matrix with six free parameters. If the resulting 9 free nonnegative parameters are regarded as equivalent to 4.5 unrestricted ones, then 50 observations, for instance, would appear to meet the rule. In addition, it can be noted that the Λ-components are actually further severely restricted by the benchmark cost matrix definition constraints. In any case, if the TMA criterion is met then the estimated model is deemed plausible. However, this test requires estimation of the $g(\omega)$-pdf as well. It would be useful for further research to develop Bayesian procedures for the insufficient data case.

8.12.2 *The Independence Assumption*

In formulating the likelihood, it was assumed that the efficiency values, η, were independently distributed across the $J \times T$ trials. Hence, a test of this would be useful to incorporate into the search procedure. However, we conjecture that lack of such independence may not necessarily negate the product formulation as a suitable fitness function for the genetic search. Future simulation studies would be useful to clarify these questions.

8.12.3 *Marginal Costs*

Unfortunately, the present method does not appear to provide direct estimates of marginal costs. That is, it would be desirable to have estimates of marginal costs per credit hour of each type for each department and period. If $\mathbf{A} = \{a_{ir}\}$ is a technically efficient benchmark cost matrix estimate then $\sum_{r=1}^{R} a_{ir}$ gives an estimate of the efficient marginal cost per credit hour of type-r for units that are efficient. Thus, this estimate should be a lower bound for marginal costs for all units.

8.13 Modification of Assumptions

8.13.1 *Limiting Drivers to Historical Ranges*

Departments may wish to maintain relative activity category, and hence driver, levels within narrower ranges than would be permitted under fully discretionary choices. We may compute from the data the observed smallest ranges, $[p_{k,h}^j, \ P_{k,h}^j]$, for which $p_{k,h}^j \leq Y_k^{jt}/Y_h^{jt} \leq P_{k,h}^j$ for all j and t, with $1 \leq k < h \leq R$. That is, these are the ranges of ratios between the various output types that have been used historically. Then the constraints $p_{k,h}^j \leq y_k^{jt}/y_h^{jt} \leq P_{k,h}^j$, which may be put in the form $\mathbf{H}^j \mathbf{y}^{jt} \leq 0$ would be added to problems **TPLP**. Alternatively, these ranges might be set as percentiles of the associated observed ratio ranges, so as to include, say, 95% of the data cases. In addition, any other known constraints for the department driver levels could be similarly added. In the genetic search, it would be necessary to add these kinds of constraints to each $\mathbf{A}$-matrix associated to a search population member.

8.13.2 *Discretionary Cost Pool Allocations*

Although not considered as affecting the example data set considered here, we briefly consider the case in which cost pools are also discretionary. This might occur when departments are allocated a budget, B^{jt}, at period-t, but may shift funds between the cost pools. We propose the following modifications of our estimation procedure for this case. Let $B^{jt} = \sum_{i=1}^{K} X_i^{jt}$ be the sum of cost pools or the total costs expended by unit-j in period t. Let x_i^{jt} be the cost pool amounts when permitted to vary over feasible choices. We assume here that the budgets are to be fully consumed. Then we may consider the modified primal problem

MTPLP(j,t):

$$\max \quad \boldsymbol{\pi}'\mathbf{y}^{jt} \tag{8.33}$$

$$\text{s.t.} \quad \mathbf{A}\mathbf{y}^{jt} - \mathbf{I}\mathbf{x}^{jt} \leq \mathbf{0}, \text{ for all } j$$

$$\text{and } t, \text{ componentwise;} \tag{8.34}$$

$$\mathbf{e}'\mathbf{x}^{jt} = B^{jt} \text{ for all } j \text{ and } t; \tag{8.35}$$

$$x_i^{jt}, \ y_r^{jt} \geq 0 \text{ for all } i, r, j \text{ and } t. \tag{8.36}$$

As before, let $\boldsymbol{\xi}^{jt}$ be the vectors of dual variables for the first K constraints (8.34). Let ξ_o^{jt} be the dual variables for the B^{jt} constraints respectively. The reader may check that after formulation of the corresponding dual problems the modified $\mathbf{MPE}(\boldsymbol{\pi}|\mathbf{A})$ model becomes:

MMPE($\boldsymbol{\pi}|\mathbf{A}$):

$$\max \quad \sum_j \sum_t \boldsymbol{\pi}'\mathbf{Y}^{jt}/B^{jt}\xi_o^{jt*} \tag{8.37}$$

$$\text{s.t. } \mathbf{A}'\boldsymbol{\xi}^{jt} = \boldsymbol{\pi} \text{ ,for all } j \text{ and } t, \tag{8.38}$$

$$\boldsymbol{\pi}'\mathbf{Y}^{jt} \leq \mathbf{X}^{jt'}\boldsymbol{\xi}^{jt*} \text{ ,for all } j \text{ and } t, \tag{8.39}$$

$$\| \boldsymbol{\pi} \|^2 \leq 1 \tag{8.40}$$

$$\pi_r, \xi_i^{jt*} \geq 0 \text{ ,for all } r, \ i, \ j, \text{ and } t. \tag{8.41}$$

8.13.3 *Department-Specific Output Priorities*

Formally, there are no difficulties in assuming differing objective coefficient vectors $\boldsymbol{\pi}^j$ for each department. In that case, the above models and estimation problems need to be modified by replacing $\boldsymbol{\pi}$ by $\boldsymbol{\pi}^j$. In addition, for the MPE model, norm constraints $\| \pi^j \|^2 \leq 1$ would be needed for each of the five departments j. Computational procedures are not appreciably more difficult. Therefore, sufficient data for the increased number of parameters would be the principal concern. For the example data of Table 8.2, there are 20 parameters in the π^j vectors; but effectively these account for 15 due to the norm constraints. With six free parameters in the cost matrix, this is a total of 21 nonnegative parameters. It should be noted that calculation of squared distance measures for the TMA criterion would be based on the modified formula

$$\omega^{jt} = \{(\ln \eta^{jt}) z^{jt*} \| \pi^j \|^{-1}\}^2 \tag{8.42}$$

8.13.4 *Department-Specific Performance Densities*

In the derivation above, it was assumed that the efficiencies η^{jt} were independently and identically distributed so that the squared distances ω^{jt} shared a common pdf $g(\omega)$. Given sufficient data, individual performance densities $g^j(\omega)$ could be estimated and used instead. However, it remains for further research to determine whether and how the TMA criterion might be applied in this case.

8.14   Conclusion and Extensions

This chapter proposes a new method for estimating cost efficiencies and benchmark costs matrices. The results provide a new tool for benchmarking studies in activity-based costing. The estimated costs provide plausible operational goals for the management of the units being compared. This method also provides efficiency measures and suggests which organizational units or time periods are more or less efficient, as well as an estimate of the degree of such inefficiency. Efficient units or periods provide benchmarks for imitation by other units or can be studied for continuous improvement possibilities. The proposed estimation approach was outlined by reference to a real data set.

Acknowldegement

The research work of this chapter is supported by the research grant committee of the Hong Kong Polytechnic University (Grant code:G-T634).

Chapter 9

Open Questions and Future Research

The Importance of Simple Questions

Vertical Density Representation has proved to yield an interesting spiral of questions, outcomes and further questions. Starting with the simple question about how performance scores are distributed, it led to useful simulation applications and connections to such diverse areas as chaos, estimation and unimodality. This reinforces the observation that an interesting question is the researcher's best friend. A colleague of the first author once stated, "I can find the density of the multivariate normal density function (which he did, in fact, by a lengthy first principles approach). But why would you want to know that ? " This book can be considered as one substantial answer to his question. It seems unlikely that this spiral has run its entire course. Thus, perhaps it may be useful to speculate on where the topic may yet lead. We believe we can best address this by stating a few of the questions we have uncovered but have not yet been able to address or carry to completion. We also mention some related works in progress. The following sections are organized around some of the key topics that were contacted in the previous chapters. These are: Benchmark Cost Matrices, Chaos, Correlation, MDR, MLE and Related Estimation Issues, Simulation, Tolstoy's Law and Unimodality. We also briefly contact the topic of probability moments, also known as frequency moments, and related integrals. This topic was mentioned in Kotz and Troutt (1996) but was not addressed earlier in this book.

9.1   Benchmark Cost Matrices

A different approach for searching the space of technically efficient benchmark cost matrices can be proposed as follows. Let $\mathbf{\Psi} = \{\psi_{ir}\}$ be a nonnegative matrix such that $\sum_i \sum_r \psi_{ir} = 1$. The ψ_{ir} may be represented by a process similar to that in Section 8.10.1. Let $E^{jt} = \{\varepsilon_{ir}\}^{jt}$ be nonnegative matrices with the same dimensions as $\mathbf{A}$. In place of (4.1), we can consider the problem

Problem $\rho^*(\mathbf{\Psi})$:

$$
\begin{aligned}
\max \quad & \rho \\
s.t. \quad & (\rho + E^{jt})\mathbf{Y}^{jt} = \mathbf{X}^{jt} \quad \text{,for all } j \text{ and } t, \\
& \rho \geq 0 \text{ and } \varepsilon_{ir}^{jt} \geq 0 \text{ ,for all } i, r, j \text{ and } t.
\end{aligned}
$$

Thus, as candidate $\mathbf{\Psi}$-matrices are generated in a genetic search process, the corresponding $\mathbf{A}$-matrix candidates are given by $\mathbf{A}^* = \mathbf{A}^*(\mathbf{\Psi}) = \rho^*\mathbf{\Psi}$. Each such $\mathbf{A}^*$-matrix clearly satisfies the following four properties:

(i) It is nonnegative.

(ii) There exist nonnegative matrices $\mathbf{A}^{jt}$, for which $\mathbf{A}^{jt}\mathbf{Y}^{jt} = \mathbf{X}^{jt}$ holds for every j and t, namely, $\mathbf{A}^{jt} = \mathbf{A}^* + E^{jt*}$.

(iii) $\mathbf{A}^* \leq \mathbf{A}^{jt} = \mathbf{A}^* + E^{jt*}$ componentwise, for every j and t.

(iv) $\mathbf{A}^*\mathbf{Y}^{jt} = \mathbf{X}^{jt}$ for at least one (j, t)-pair. Otherwise, the optimality of ρ^* can be contradicted. Thus, each such $\mathbf{A}^*$-matrix must be technically efficient.

The relative merits of this approach and that proposed in Chapter 8, or other possible approaches are worthy of further study. Condition (iii) might be described as providing a *strong* technically efficient benchmark cost matrix. We conjecture that the process described in Chapter 8 does not necessarily generate cost matrices with this property.

9.2   Chaos

As of the time of this writing, no results on cycling had been published for the sharkfin class of chaos generators discussed in Chapter 5. However, Woodard, Chen and Madey (2002) have recently obtained some computational results on the sharkfin generator. They proposed a cycle detection method similar to those used in Sedgewick and Szymanski, (1978), Su (1998) and Fich (1983). They report that cycles appear after 2^{26} iterations. This seems relatively short, considering that the internal program representation for the iterates was 16 bytes (128 bits). Thus, further research is needed to improve the cycle lengths of generators of this kind. However, with the use of such tests for cycle completion, a restart strategy could be used to extend the effective cycle length for the sharkfin generators. That is, a switch to a different α-value could be employed as soon as each cycle is revealed. As an alternative, a fixed number of recursions, for example, 10^6, might be used with similar restarts.

Work on examining correlations between different sharkfin generators is yet to be done. Performance of the two-stage sharkfin, in which both the initial iterate and α-parameter are recursively updated, has yet to be tested for the uniform distribution property of orbit values.

It should prove very useful if a characterization of uniformly distributed chaotic orbits on the unit hypercube could be developed similar to Theorems 5.1 and 5.2. The basic intuition of that result will hold equally well if the recursions are based on vectors instead, as in $\mathbf{x}_{n+1} = V(\mathbf{x}_n)$ with V a vector-valued function. However, it is not clear whether an analog of the $A'(v)$ construct exists for that context or whether an entirely new approach will need to be devised.

9.3 Correlation

The results we have obtained on decomposition of correlation in Section 2.3 depended only on the most basic VDR considerations of Theorem 1.1, combined with a simple scheme to represent dependency in being above or below the mean. More advanced VDR results such as Theorem 1.2 and Theorem 1.5 may likely yield further interesting analyses, as may developments that follow Type II VDR (Chapter 4) , KDR (Chapter 6), and the discussion in Troutt and Pang (1997). In particular, we believe it will be useful to better understand the role of the conditional pdf $h(\mathbf{x}|V(\mathbf{x}) = v)$ in dependency of random vectors.

In our univariate decomposition of correlation, we obtained two components. The first is the vertical component, which is the correlation of the V_1 and V_2 random variables associated to two other random variables

X_1 and X_2 by way of their pdfs. That is, $V_i = f(X_i)$. We note that in the multivariate setting, two random vectors may have this same type of vertical correlation component. For example, in meteorology, wind vectors at two different locations might, in principle, be correlated to different degrees in magnitudes, whether or not one defines a correlation of directions. We conjecture that the vertical correlation in that setting is related to the correlation in magnitudes.

We feel that use of VDR results with the topic area of copulas will be fruitful in the future and that VDR methods may provide an alternative to their use. Copulas are functions that are useful for modeling dependencies when marginal distributions are known but the true joint density is difficult to specify. Background theory on copulas is presented in Nelson (1995) and in Schweizer (1991) and recent applications are discussed in Berardi, Patuwo and Hu (2002), Clemen and Reilly (1999), Yi and Bier (1998), Jouini and Clemen (1996). We conjecture that dependencies may be modeled by the appropriate choices of the $V(\mathbf{x})$ and $h(\mathbf{x}|V(\mathbf{x}) = v)$ functions of VDR.

9.4　MDR, MLE and Related Estimation Issues

The MDR estimation principle requires minimizing the sum of the shortfalls or distances from the target or ideal. Thus, if $\delta_t(\boldsymbol{\theta})$ is the shortfall for trial-t as a function of the unknown parameter vector $\boldsymbol{\theta}$ then the MDR estimate is given by the admissible vector $\boldsymbol{\theta}^*$ which minimizes $\sum_t \delta_t(\boldsymbol{\theta})$. Decision-makers can hardly argue with this in the sense that if asked the question, "Have you attempted to minimize $\sum_t \delta_t(\boldsymbol{\theta})$?", then they can surely agree. However, they should be expected to be equally agreeable if the objective is changed to any monotone increasing function of the $\delta_t(\boldsymbol{\theta})$, for example, $\sum_t \exp\{\delta_t(\boldsymbol{\theta})\}$. In fact, they should be able to agree that they have attempted to minimize any and all such objectives. However, one is not necessarily assured that all these criteria will lead to the same estimate and a resolution of this question awaits further research. One possibility is to select from such a set of criteria one that has some additional normative feature. For example, suppose it were known in advance that the $\delta_t(\mu^*)$ are distributed according the exponential distribution with pdf $g(\delta) = \mu e^{-\mu\delta}, \delta \geq 0$. Then it is easy to see that minimization of $\sum_t \delta_t(\mu)$ yields a MLE estimate of μ^*. Similarly, if it were known in advance that the

$\delta_t(\mu^*)$ are distributed according to the half-normal density, then minimization of $\sum_t \{\delta_t(\mu)\}^2$ would provide a MLE estimate of μ^*. Thus, in general, one might perform a search over possible monotone transformations of the $\delta_t(\boldsymbol{\theta})$ in hopes of finding one for which the resulting distribution of the $\delta_t(\boldsymbol{\theta}^*)$ has the appropriate exponential density that implies $\boldsymbol{\theta}^*$ is a MLE estimate.

However, what appears to be the usual outcome is that the resulting distribution of the $\delta_t(\boldsymbol{\theta}^*)$ will be one for which the mode is strictly greater than zero. For example, in the benchmark cost estimation application of Chapter 8, the best-fitting distribution was the gamma with a shape parameter associated with a positive mode. Here lies an intriguing issue for further research. Why should a decision-maker make efforts at achieving a distribution of the $\delta_t(\boldsymbol{\theta}^*)$, which has a positive mode? That appears to be what would occur if the decision-maker aims at maximizing the likelihood rather than minimizing $\sum_t \delta_t(\boldsymbol{\theta}^*)$. We conjecture that the MDR objective will prove to be a better representation of purposeful behavior than is the MLE criterion, whether or not the resulting distribution of the shortfalls has positive mode.

These results also raise interesting questions about MLE estimation itself. For example, it would be possible, in principle, to consider MLE estimation based on $g(v)$. What are the properties of such estimates and when might they be more useful?

The likelihood function for i.i.d. assumptions is itself a joint pdf function. Its own $g(v)$ might be further examined. As a random variable in its own right, the likelihood varies from sample to sample. Despite its many desirable properties, MLE can be regarded as counterintuitive from this perspective. That is, one might argue that rather than a sample having maximum likelihood, perhaps it is more reasonable to assume that it should have *mean likelihood*. Thus, under this view, parameter estimates would be chosen so that the "observed" likelihood of the sample coincides with the average likelihood for the parameter estimates selected. Some preliminary work with this notion for the univariate normal density suggests that such a principle does not produce unique estimates of the mean and variance. Rather, it appears to produce an implicit relationship between their estimates. Such an approach may therefore be useful when one or more additional relationships between the parameters are known in advance.

We next observe that the MDE/MPE/MDR class of estimators may be regarded as Decision Theory *in reverse*. To see this, we note that in elementary Decision Theory (DT), (see for example, Clemen, 1996), the decision-maker knows the possible states of nature, typically also their probabilities and finally the payoffs that result from each state and decision choice or action. Based on that information, a decision, mixed strategy or decision strategy is undertaken. One well-known criterion is to minimize regret or expected regret, where regret is the difference between optimal payoff for a state and the actual payoff for the decision taken. The unknown parameters to be estimated in these methods, such as costs, are analogous to the states of nature in the DT context. In MDR estimation, we essentially assume that expected regret has been minimized and search for the estimate consistent with that assumption. This suggests that there may be a kind of reliability process or measure involved. Namely, it suggests that decision-making performance is likely to be better in the neighborhood of the minimum regret decision. That is, smallness of the mean regret measure should be associated with closeness to the optimal decision. Such assumptions might be empirically tested with real decision makers for future research. In addition, the dart-throwing game is primarily a physical activity while real world decision-making is presumably entirely mental. It remains for research to determine if there are fundamental differences in performance variability between physical and mental tasks.

The analogy of decision performance to throwing darts at a bull's-eye target was also the basis of the normal-like-or-better criterion of Chapter 8. The distances from the target possibly do not constitute a truly iid sample from one and the same distribution. In fact, one could hope that several kinds of continual improvements may be operating in such a process, namely: systematic reductions in bias and variance, respectively, as well as improving steepness of the density at its mode. In addition, one might further assume a mixture of white noise and performance shortfalls along the lines of stochastic frontier estimation mentioned in Chapter 8. Learning curve models (see for example, Zangwill and Kantor, 1998) may be suitable for modeling the improvements of such parameters. However, a learning curve model introduces a further parameter called the learning rate. We believe the Markov Chain Monte Carlo estimation technique (see for example, Pang *et al.*, 2001 and 2002) may be useful for such complex estimation tasks. Despite these concerns, the iid assumption may still be a suitable approximation for small samples, particularly if all the parameter improvement rates are small enough.

9.5 Probability (Frequency) Moments and Related Integrals

Sichel (1947, 1949) defined the rth frequency (probability) moment of a pdf $f(x)$ on $\Re$, as follows:

$$\Omega_r = \int_{\Re} (f(x))^r \mathrm{d}x \tag{9.1}$$

where evidently $\Omega_1 = 1$. See also Kotz and Johnson (1989), pages 122-123. If $[0, m]$ is the range of $V(x) = f(x)$ for $x \in \Re$ and $g(v)$ is the vertical density, then we have

$$
\begin{aligned}
\Omega_r &= \int_{\Re} (f(x))^r dx & (9.2) \\
&= \int_{\Re} (f(x))^{r-1} f(x) dx & (9.3) \\
&= \mathrm{E}_X((f(x))^{r-1}) & (9.4) \\
&= \mathrm{E}_V(v^{r-1}) & (9.5) \\
&= \int_{[0,m]} v^{r-1} g(v) dv & (9.6)
\end{aligned}
$$

Thus, the rth probability moment of $f(x)$ is the $(r-1)^{\text{st}}$ moment of $g(v)$. In addition, recalling that the dual densities $A(v)$ and $g(v)$ are related by $g(v) = -vA'(v)$ from Theorem 1.1, and denoting by E_A the expectation of the random variable with pdf $A(v)$, the reader may verify that

$$\mathrm{E}_A(v^{r-1}) = r^{-1} \mathrm{E}_V(v^{r-1}) \tag{9.7}$$

More generally, for suitable conditions on $\varphi(\cdot)$, we expect the validity of the following transformation formula:

$$\int_{\Re} \varphi(f(x)) f(x) dx = \int_{[0,m]} \varphi(v) g(v) dv \tag{9.8}$$

The potential usefulness and extensions of these relationships have so far not been explored.

9.6   Simulation

Simulation from a given $g(v)$-pdf can be accomplished by simulating from $f(\mathbf{x})$ and then transforming the results via $v = f(\mathbf{x})$. More generally, to simulate from a new pdf, we might consider it as a possible $g(v)$-pdf and then seek the corresponding $f(\mathbf{x})$-pdf. The latter pdf might be obtained by first solving the differential equation, $A'(v) = -v^{-1}g(v)$, which results from (1.2). An illustration of solving this equation was given in Section 2.2, which also showed how a symmetric univariate $f(x)$ can be constructed from knowledge of $A(v)$ using the strata shift idea. The same idea should be adaptable to the multivariate setting given the appropriate use of the strata shift idea for that context.

Theorem 1.5 raises an interesting question in connection with simulation and what we may call *extended randomization*. We have

$$f(\mathbf{x}) = g(v)\|V(\mathbf{x})\|h(\mathbf{x}|V = v),$$

for $\mathbf{x} \in \{V = v\}$ and realizations of the associated random variable can be obtained by the following steps: (i) select v according to $g(v)$ and then (ii) select $\mathbf{x}$ on $\{V = v\}$ according to $h(\mathbf{x}|V = v)$. However, consider the following modified process. Suppose the contours, $\{V = v\}$ are circles in $\Re^2$ and that $h(\mathbf{x}|V(x) = v)$ is the wrapped normal density, say, with mode $\mu(v) \in \{V = v\}$ and a fixed variance σ^2 (see, for example, Mardia, 1972 or Fisher *et al.*, 1987). One might select $\mu(v)$ itself according to a uniform distribution, for example, on $\{V = v\}$, a process that can be called extended randomization. Thus more generally, $h(\cdot)$ may depend on a parameter vector $\boldsymbol{\theta}$ as in $h(\mathbf{x}|V(\mathbf{x}) = v; \boldsymbol{\theta})$ which could have its own prior pdf $p(\boldsymbol{\theta})$. This case has not yet been explored. Simulation of such a process appears to present no particular difficulties. However, it is not clear whether the $f(\mathbf{x})$-formula of Theorem 1.5 makes any sense in such cases.

These considerations provide a further avenue for density modeling. In Chapter 1, a number of univariate density examples were constructed by combining the $g(v)$-pdf from one class and the $V(\mathbf{x})$-function from another class. The formula in Theorem 1.5 enables a further degree of freedom in such a construction by permitting the possibility of choosing the $h(\mathbf{x}|V(\mathbf{x}) = v)$ conditional or contour density from a third class.

9.7 Tolstoy's Law

The Tolstoy's Law condition, (6.5), of Theorem 6.4, proved useful in developing the target-mode agreement criterion for MDR estimates. The original motivation for Theorem 1.5 was to obtain a result like Tolstoy's Law when the contour density $h(x|V = v)$ is not uniform, which should provide a broadening of applicability for those results. That task remains to be completed. Intuition suggests that the mode coincidence property of Tolstoy's Law should still hold if all the contour densities are themselves unimodal and the path of their modes, as a function of v, has sufficient smoothness. However in that case, the simplicity of the $A(v)$-construct is lost in Theorem 6.4. Nevertheless, it may be possible to start with the more general relation of Theorem 1.5, and obtain a condition similar to (6.5) in some special cases. We conjecture that one case of interest will occur when $\| \nabla V(\mathbf{x}) \|$ is a function of v alone.

9.8 Unimodality

It should be noted that in Section 6.4, we have not obtained a proof of Khintchine's Unimodality Theorem (KUT), *per se*, but rather a proof of the formula (6.25) of that theorem. In effect, we have proved a limited version of KUT on the half-interval. However, we believe that proof of the general form of the KUT may follow by an appropriate use of the strata shift idea of Section 6.5. In particular, we conjecture that any pdf that is unimodal about zero is equivalent to a monotone decreasing pdf on the half-interval in the strata shift sense.

An intuitive explanation of KUT on the half-interval can be given with the help of Fig. 6.2. We have $h(x|z) = z^{-1}$ if $z \geq x$ and $h(x|z) = 0$, otherwise. Thus, $f(x) = \int_{z \geq 0} h(x|z)g(v)dv = \int_{z > x} z^{-1}g(z)dz$. Assuming $g(z) > 0$ on $(0, \infty)$ and $y \geq x$, we have $f(x) = \int_{z > x} z^{-1}g(z)dz = f(y) + \int_{x \leq z \leq y} z^{-1}g(z)dz > f(y)$. Thus, $f(x)$ must be monotone strictly decreasing on $(0, \infty)$. Of course, further instances of such representations could be similarly constructed by choosing the $h(x|z)$ conditional pdf as other than uniform on $[0, z]$. We note there is a similarity between the Type II VDR (see Sections 3.1 and 4.5) and the KDR of Section 6.4. We conjecture that the difference is primarily one of strata shift equivalence.

We believe that many other interesting relationships among the various associated densities on the half-interval may exist beyond those obtained in Section 6.4.

Bibliography

Ackoff R. L. (1962). *Scientific Method.* Wiley, New York.

Ahrens J. H. and Dieter U. (1972). Computer methods for sampling from the exponential and normal distributions, *Comm. ACM* , **15**, 872-882.

Ahrens J. H. and Dieter U. (1974). Computer methods for sampling from gamma, beta, Poisson, and binomial distributions, *Computing*, **12**, 223-246.

Ahrens J. H. and Dieter U. (1988). Efficient table-free sampling method for the exponential, Cauchy and normal distributions, *Comm. ACM* , **31**, 1330-1337.

Ahrens J. H. and Dieter U. (1991). A convenient sampling method with bounded computation times for Poisson distributions, *The Frontiers of Statistical Computation, Simulation, & Modeling*, (edited by P. R. Nelson, E. J. Dudewicz, A. Öztürk, and E. C. van der Meulen) American Science Press, Columbus, Ohio, 137-149. **12**, 223-246.

Aigner D. J. and Chu S. F. (1968). On estimating the industry production function. *American Economic Review.* **58**, 826-839.

Aigner D. J., Lovell C. A. K. and Schmidt P. (1977). Formulation and estimation of stochastic frontier production function models. *Journal of Econometrics.* **6**, 21-37.

Alsalem A. S., Sharma S., and Troutt, M.D. (1997). Fairness Measures and Importance Weights for Allocating Quotas to OPEC Member Countries. *The Energy Journal*, **18**, 2, April, 1-21.

Anderson S. L. (1990). Random number generators on vector supercomputers and other advanced architectures. *SIAM Rev.* **32**, 221-251.

Atkinson A. C. and Pearce M. C. (1976). The Computer Generation of Beta, Gamma and Normal Random Variables. *J. Roy. Stat. Soc.*, **A139**, 431-448.

Avriel M. (1976). *Nonlinear Programming Analysis and Methods.* Prentice-Hall, Inc. Englewood Cliffs, New Jersey.

Banker R. D. and Johnston H. H. (1993). An empirical study of cost drivers in the U.S airline industry. *The Accounting Review.* **68**, (3), 576-601.

Bartlett J. (1992) *Familiar Quotations*, Sixteenth Edition, Little Brown and Company, Boston.

Bartlett M. S. (1934). The vector representation of a sample. *Proc. Camb. Phil. Soc.* **30**, 327-340.

Bazaraa M. S., Jarvis J. J. and Sherali H. D. (1990). *Linear Programming and Network Flows*. 2nd Ed. John Wiley & Sons, New York.

Bazaraa M. S., Sherali H. D. and Shetty C. M. (1993). *Nonlinear Programming: Theory and Algorithms*. 2nd Ed. John Wiley & Sons, Inc. New York.

Berardi V., Patuwo B. E. and Hu M. Y. (2002). Detailed Procedure for Using Copulas to Classify E-business Data ? *Working paper*, Department of Management & Information Systems, Kent State University, Kent, Ohio 44242.

Bhattacharyya S. and Troutt M. D. (2000). Crossover in Probability Spaces. *Proceedings of the Genetic and Evolutionary Computation Conference (GECCO-2000)*, D. Whitley, D.Goldberg, E. Cantu-Paz, L. Spector, I. Parmee, H. G. Beyer (Eds.), Las Vegas, July 10-12, 120–127, Morgan Kaufmann Pub. Co.

Bhattacharyya S. and Troutt M. D. (2001). Genetic Search over Probability Spaces. *European Journal of Operational Research.* **144**, (2) (January), 333-347.

Bhavsar V. C. and Isaac J. R. (1987). Design and analysis of parallel Monte Carlo algorithms,. *SIAM J. Sci. Statist Comput.* **8**, s73-s95.

Berger M. (1987). *Geometry II*. Springer-Verlag, Berlin.

BestFit (1995). *User's guide*. Newfield, NY: Palisade Corporation.

Bickell P. J. and Doksum K. A. (1977). *Mathematical statistics: Basic ideas and selected topics*. San Francisco: Holden Day, Inc.

Bowman E. H. (1963). Consistency and Optimality in Managerial Decision Making. *Management Science.* **9**, (January) 310-321.

Bowman R. L. (1995). Evaluating Pseudo-Random Number Generators. *Comput. & Graphics.* **19**, No. 2, 315-324.

Box G. E. P. and Muller M. E. (1958). A note on the Generation of Random Normal Deviates, *Ann. Math. Statist.*, **29**, 610-611.

Brogan W. L. (1991). *Modern Control Theory*. 3rd ed. Prentice-Hall, Inc., Englewood Cliffs, New Jersey.

Brown R. G. (1961). Use of the Carrying Charge to Control Cycle Stocks, *APICS (American Production and Inventory Control Society) Quart. Bull.* **2**, (3) (July), 29-46.

Brown R. G. (1967). *Decision Rules For Inventory Management*. Holt, Rinehart and Whinston, New York.

Büeler B., Enge A. and Fukuda K. (1998). Exact volume computation for convex polytopes: A practical study. In G. Kalai and G. Ziegler, editors, Polytopes - Combinatorics and Computation. DMV-Seminars. Birkhauser Verlag.

Bullard J. and Butler A. (1993). Nonlinearity and Chaos in Economic Models: Implications for Policy Decisions. *The Economic Journal* Vol. **103**, No. 419, 849-867.

Butler A. (1990). A methodological approach to chaos: are economists missing the point? Federal Reserve Bank of St. Louis Review. Vol. **72** (March/April), 36-48.

Buxey G. (1995). A managerial perspective on aggregate planning. *International Journal of Production Economics.* **41**, 127-133.

Charnes A., Cooper W. W., Lewin A. and Seiford L. M. (1994). *Data envelopment analysis: Theory, methodology, and applications.* Boston: Kluwer Academic Publishers.

Chen H. C. and Asau Y. (1974). On generating random variates from an empirical distribution, *AIIE Transactions,* **6**, 163-166.

Chen M. H., Shao Q. M. and Ibrahim J. G. (2000). *Monte Carlo Methods in Bayesian Computation.* Springer, Berlin.

Cheng R. C. H. and Feast G. M. (1979). Some simple gamma variate generators, *Applied Statistics,* **28**, 290-295.

Chmielewski M. A. (1981). Elliptically symmetric distributions:A review and bibliography. *Inter. Statist. Review.* **49**, 67-74.

Clemen R. T. and Reilly T. (1999). Correlations and Copulas for Decision and Risk Analysis. *Management Science.* **45**, 2, 208-224.

Clemen R.T. and Winkler R.L. (1985). Limits for the Precision and Value of information from Dependent Sources. *Oper. Res.,* **33**, 427-442.

Clemen R.T. and Winkler R.L. (1990). Unanimity and Compromise Among Probability Forecasters. *Management Sci.,* **36**, 767-779.

Clemen R.T. and Winkler R.L. (1993). Aggregating Point Estimates: A flexible Modeling Approach. *Management Sci.,* **39** , 501-515.

Coelli T., Prasada Rao D. S. and Battese G. E. (1998). *An Introduction to Efficiency and Productivity Analysis.* MA:Kluwer Academic, Boston.

Cohen J. and Hickey T. (1979). Two Algorithms for Determining Volumes of Convex Polyhedra. *Journal of the Association for Computing Machinery.* **26**, (3), 401-414.

Crutchfield J., Farmer J. and Huberman B. (1982). Fluctuations and simple chaotic dynamics. *Physics Reports.* Vol. **92**, 45-82.

Deusing E. C. (1977). *Polyhedral Convex Sets and the Economic Analysis of Production.* Unpublished Ph.D. dissertation, Department of Economics, University of North Carolina at Chapel.

Devaney R. L. (1989). *An Introduction to Chaotic Dynamical Systems.* Addison-Wesley, Redwood City.

Devroye L. (1986). Non-Uniform Random Variate Generation. *Springer-Verlag.*

Devroye L. (1987). A simple generator for discrete log-concave distributions. *Computing,* **39**, 1, 87-91.

Dharmadhikari S. and Joag-dev K. (1988). *Unimodality, Convexity, and Applications.* Academic Press, Inc. Boston.

Dopuch N. and Gupta M. (1997). Estimation of benchmark performance standards: An application to public school expenditures. *Journal of Accountancy and Economics.* **23**, 141-161.

Dubois F. and Oliff M. D. (1991). Aggregate Production-planning in Practice. *Production and Inventory Management Journal.* **32**, (3), 26-30.

Dyson R. G. and Thanassoulis E. (1988). Reducing weight flexibility in data envelopment analysis. *Journal of the Operational Research Society.* Vol. **39**, No. 6, 563-576.

Eddy W. F. (1990). Random number generators for parallel processors. *J. Comp. Appl. Math.* **31**, 63-71.

Eilon S. (1975). Five Approaches to Aggregate Production Planning. *AIIE Transactions.* **7**, (2), (June) 118-131.

Einhorn H. J. (1974). Expert Judgement: Some Necessary Conditions and an Example. *Journal of Applied Psychology.* Vol. **59**, No. 5, 562-571.

Elnathan D. and Kim O. (1995). Partner selection and group formation in cooperative benchmarking. *Journal of Accounting and Economics.* **19**, 345-364.

Elnathan D., Lin T. W. and Young S. M. (1996). Benchmarking and management accounting: A framework for research. *Journal of Management Accounting Research.* **8**, 37-54.

Falk J. E. and Palocsay S. W. (1992). Optimizing the Sum of Linear Fractional Functions. *Recent Advances in Global Optimization,* C. A. Floudas and P. M. Pardalos, (eds.), 221-258.

Fang K. T., Kotz S. and Ng K. W. (1990). *Symmetric multivariate and related distributions.* Chapman and Hall, New York.

Fang K.T. and Wang Y. (1994) *Number-theoretic Methods in Statistics.* Chapman and Hall, London.

Fang K. T., Yang Z. H. and Kotz S. (2001). Generation of multivariate Distributions by Vertical Density Representation. *Statistics,* **35**, 281-293.

Feichtinger G. (1996). Chaos Theory in Operations Research. *Int. Trans. Opl. Res.* Vol. **3**, No. 1, 23-36.

Feller W. (1971). *An Introduction to Probability Theory and Its Applications.* 2nd ed., Vol. **2**. Wiley, New York.

Fich F. E. (1983). Lower Bounds for the Cycle Detection Problem. *J. Comput. System Sci.* **26**, no. 3, 392-409.

Fisher N. I., Lewis T. and Embleton B. J. (1987). *Statistical Analysis of Spherical Data.* Cambridge University Press, Cambridge.

Fleming W. (1977). *Functions of several variables.* 2nd ed., New York: Springer-Verlag.

Gamerman D. (1997). *Markov chain Monte Carlo: Stochastic Simulation for Bayesian Inference.* Chapman and Hall, London.

Ganest C. and Zidek J.V. (1986). Combining Probability Distributions, with Discussion and Annotated Bibliography. *Statistical Sci.,* **1** 114-147.

Gentle J. E. (1998). *Random Number Generation and Monte Carlo Methods,* Springer, New York.

Gilks W. R., Richardson S. and Spiegelhalter D. J. (1996). *Markov Chain Monte Carlo in Practice.* Chapman and Hall, London

Gilks W. R. and Wild P. (1992). Adaptive Rejection Sampling for Gibbs Sampling,

Appl. Statist. **41**, 2, 337-348.

Gilbert S. M. (2000). Coordination of Pricing and Multiple-Period Production Across Multiple Constant Priced Goods. *Management Science.* **46**, (12) 1602-1616.

Glover F. and Woolsey E. (1974). Converting the 0-1 Polynomial Programming Problem to a 0-1 Linear Program. *Operations Research.* **22**, 180-182.

Goodman I. R. and Kotz S. (1981). Hazard Rates Based on Isoprobability Contours. *Statistical Distributions in Scientific Work,* **5**, 289-309. Taillie C. and Patil G. P. eds.; Reidel, Dordrecht, Netherlands.

Gray P. and Watson H. J. (1998). *Decision Support in the Data Warehouse.* Prentice-Hall PTR, Inc., Englewood Cliffs, NH.

Green W. H. (1990). A gamma-distributed stochastic frontier model. *Journal of Econometrics.* **46**, 141-163.

Harrell F. E. (2001). *Regression Modeling Strategies with Applications to Linear Models, Logistic Regression, and Survival Analysis.* Springer, New York.

Hartman P. and Wintner A. (1940). On the spherical approach to the normal distribution law. *Am. J. Maths.* **62**, 759-779.

Hastings W.K. (1970). Monte Carlo Sampling Methods Using Markov Chains and Their Applications. *Biometrika,* **57**, 97-109.

Hax A. C. and Candea R. (1984). *Production and Inventory Management.* Prentice- Hall, Inc., Englewood Cliffs, NH.

Holt C. C., Modigliani F. and Simon H. A. (1955). A Linear Decision Rule For Production and Employment Scheduling. *Management Science,* (October) 1-30.

Holt C. C., Modigliani F. and Muth J. F. (1956). Derivation of A Linear Decision Rule For Production and Employment. *Management Science,* (January) 159-177.

Horngren C. T., Foster G. and Datar S. M. (2000). *Cost accounting: A managerial emphasis.* 10th ed. Upper Saddle River, New Jersey: Prentice-Hall Inc.

Hörmann, W. (1994), A universal generator for discrete log-concave distributions. *Computing,* **52**, 1, 89-96.

Hörmann W. and Derflinger G. (1997). An automatic generator for a large class of unimodal discrete distributions. In *EMS97, Gaining Competitive Advantage Through Simulation Technologies,* A. R. Kaylan and A. Lehmann, Eds. 139-144.

IMSL (1991). FORTRAN Subroutines for Statistical Ananlysis. IMSL Inc., Texas, USA.

Isaacs R. (1965). *Differential Games.* John Wiley and Sons, New York.

Johnson N. L. and Kotz, S (1987). *Continuous Univariate Distributions-1 & 2.* John Wiley & Sons, New York.

Johnson, N. L. and Kotz, S. (1999). Non-smooth Sailing or Triangular Distributions Revisited after Some 50 Years. *The Statistician,* **48**, 2, 179-187.

Jouini M. and Clemen R. T. (1996). Copula Models for Aggregating Expert Opinions. *Operations Research.* **44**, 3, 444-457.

Kemp A. W. (1990). Patchwork rejection algorithms. *J.Comput. Appl.Math.* **31**, 1, 127-131.

Kellison S. G. (1975). *Fundamentals of Numerical Analysis.* Richard D. Irwin, Inc. Homewood, Illinois.

Khinchine A. Y. (1938). Tomskii Univ. Nauch. Issled. Inst. Mat.-Mekh., Izv., 2, 1-7.

Kindermann A. J. and Monahan J. F. (1977). Computer generation of random variables using the ratio of uniform deviates. *ACM Trans.Math. Software*, **3**, 257-260.

Kindermann A. J. and Monahan J. F. (1980). New methods for generating Student's *t* and gamma variables. *Computing*, **25**, 2369-377.

Kinderman A. J. and Ramage J. G. (1976). Computer generation of normal random variables. *Journal of The American Statistical Society*, **71**, 893-896.

Kleinbaum D. G., Kupper L. L., Muller K. E. and Nizam A. (1997). *Applied Regression Analysis and Other Multivariable Methods.* 3rd ed. Duxbury Press, Brooks/Cole Publishing Company, Pacific Grove. California.

Knuth D. E. (1981). *The Art of Computer Programming.* Vol. **2**, 2nd ed. Addison-Wesley, Reading, MA.

Konno H. and Kondo T. (1997). Iterative Chaotic Map as Random Number Generator. Technical Note. *Ann. Nucl. Energy.* Vol. **24**, No. 14, 1183-1188.

Kotz S., Fang K. T. and Liang Jia-Juan (1997). On multivariate vertical density representation and its application to random number generation. *Statistics*, **30**, 163-180.

Kotz S. and Johnson N. L. (1989). Editors-in-chief. Article: Khinchins Unimodality Theorem, 77-78. *Encyclopedia of Statistical Sciences, Supplement Volume.* John Wiley & Sons, New York.

Kotz S. and Troutt M. D. (1996). On vertical density representation and ordering of distributions. *Statistics*, **28**, 241-247.

Kozubowski T. J. (2002). On the vertical density of the multivariate exponential power distribution. *Statistics*, **36**, 219-221. York.

Kunreuther H. (1969). Extensions of Bowman's Theory on Managerial Decision-Making. *Management Science.* **16**, (8) (April) B-415 through B-439.

Lau A. and Lau H. (1981). A Comment on Shih's General Decision Model for CVP Analysis. *The Accoutning Review.* Vol. **56**, (October), 980-283.

Lau A. and Lau H. (1983). Towards a Theory of Stochastic Exit Value. *Accounting and Business Research.* Vol. **14**, (43), 21-28.

Law A. M. and Kelton W. D. (1982) *Simulation Modeling and Analysis*, McGraw-Hill, New York.

Lawrence J. (1991). Polytope volume computation. *Mathematics of Computation.* **57**, 196, October, 259-271.

Lee W. B. and Khumawala B. M. (1974). Simulation Testing of Aggregate Production Models in an Implementation Methodology. *Management Science.* **20**, (6), 903-911.

Li T. and Yorke J. A. (1975). Period Three Implies Chaos. *American Mathematical*

Monthly. Vol. **82**, 985-992.

Madansky A. (1988). *Prescriptions for working statisticians*. Springer-Verlag, New York.

Marsaglia G. (1963). Generating discrete random variables in a computer. *Communications of the ACM*, **6**, 37-38.

Marsaglia G. (1977). The squeeze method for generating gamma variates, *Comput. and Math. with Appl.*, **3**, 321-325.

Marsaglia G. (1968). Random Numbers Fall Mainly in the Planes. *Natl. Acad. Sci. Proc.* **61**, 25-28.

Marsaglia G. (1984). The exact-approximation method for generating random variates in a computer, *J. Amer. Statist. Assoc.*, **79**, 218-221.

Marsaglia G. and Bray T. A. (1964). A convenient method for generating normal variables, *SIAM Review*, **6**, 260-264.

Mattheis T. H. and Rubin D. S. (1980). A Survey and Comparison of Methods for Finding all Vertices of Convex Polyhedral Sets. *Mathematics of Operations Research*. **7**, (2), 167-185.

McCaffrey D., Ellner S., Gallant A. and Nychka D. (1992). Estimating the Lyapunov exponent of a chaotic system with nonparametric regression. *Journal of the American Statistical Association*. Vol. **87**, No. 419, 682-695.

Menon S. and Sharda R. (1999). Digging Deeper - Data Mining update: new modes to pursue old objectives. *OR/MS TODAY*. June. 26-29.

Metropolis N., Rosenbluth A. W., Rosenbluth M. N., Teller A. H. and Teller E. (1953). Equations of state calculations by fast computing machines. *J. Chem. Phys.*, **21**, 1087-1091.

Morrison D. F. (1976). *Multivariate Statistical Methods*. 2nd ed. McGraw-Hill, New York.

Moskowitz H. and Miller J. G. (1975). Information and Decision Systems for Production-planning. *Management Science*. **22**, (3) 359-370.

Muirhead R. J. (1982). *Aspects of Multivariate Statistical Theory*. John Wiley & Sons, New York.

Nam S. J. and Logendran R. (1992). Aggregate production-planning - A survey of models and methodologies. *European Journal of Operational Research*. **61**, 255-272.

Nelson R. B. (1995). Copulas, characterization, correlation, and counterexamples. *Mathematics Magazine*. **68**, 193-198.

Neter J., Wasserman W. and Kutner M. H. (1985). *Applied linear statistical models*. 2nd ed. Homewood, Illinois: Richard E. Irwin, Inc.

Niederreiter H. (1994). Pseudorandom vector generation by the inversive method. *ACM Trans. Modeling and Computer Simulation*. **4**, 191-192.

Niederreiter H. (1995). Pseudorandom vector generation by the multiple-recursive matrix method. *Math. Comp.* **64**, 279-294.

Norman J. E. and Canon L. E. (1972). A computer program for generation of random variables from any discrete distribution, *Journal of Statistical Computation and Simulation*, **1**, 331-348.

Ornstein D.S. (1995). In what Sense can a Deterministic System be Random ? *Chaos, Solitons & Fractals.* Vol. **5**, No. 2. 139-141.

Ott E. (1993). *Chaos in dynamical systems.* Cambridge University Press, Cambridge

Pang W. K., Yang Z. H., Hou S. H. and Troutt M. D. (2001). Some Further Results of Multivariate Vertical Density Representation and Its Application. *Statistics,* **35**, 463-477.

Pang W. K., Yang Z. H., Hou S. H. and Leung P. K. (2002). Non-uniform random Variate Generation by Vertical Strip Method with Given Density. *European Journal of Operational Research,* **142**, 595-609.

Pearson C. E. (1974). *Handbook of Applied Mathematics.* Van Hostrand Reinhold Company, New York.

Peterson R. and Silver E. A. (1979). *Decision Systems for Inventory Management and Production-planning.* John Wiley & Sons, New York.

Quesada I. and Grossman I. E. (1995). A Global Optimization Algorithm for Linear Fractional and Bilinear Programs. *Journal of Global Optimization.* **6**, 39-76.

Rajagopalan S. and Swaminathan J. M. (2001). A Coordinated Production-planning Model with Capacity Expansion and Inventory Management. *Management Science.* **47**, (11), November, 1562-1580.

Rajasekaran S. and Ross K. W. (1993). Fast algorithms for generating discrete random variates with changing distributions, *ACM Transactions on Modeling and Computer Simulation,* **3**, 1-19.

Reza F. M. (1961). *An Introduction to Information Theory.* McGraw-Hill Book Company, Incorporated, New York.

Riggle C. and Madey G. (1997). An analysis of the impact of chaotic dynamics on management information flow models. *European Journal of Operational Research.* **103**, 242-254.

Ritter C. and Léopold S. (1997). Pitfalls of normal-gamma stochastic frontier models. *Journal of Productivity Analysis.* **8**, 167-182.

Robinson A. G. and Dilts D. M. (1999). OR & ERP - A Match for the New Millennium. *OR/MS TODAY.* June. 30-35.

Saaty T. L. (1980). *The Analytic Hierarchy Process,* MaGraw Hill, New York.

Schactman R. H. (1974). Generation of the Admissible Boundary of a Convex Polytope. *Operations Research.* **22**, 151-159.

Shih W. (1979). A General Decision Model for Cost-Volume-Profit Analysis Under Uncertainty. *The Accounting Review.* Vol. **54**, (October), 687-706.

Schroeder R. G. (1993). *Operations Management: Decision Making in the Operations Function.* 4th ed. McGraw-Hill, Inc., New York.

Schweizer B. (1991). Thirty years of copulas. G. DallAglio, S. Kotz, G. Salinetti, eds. *Advances in Probability Distributions with Given Marginals,* Kluwer, Dordrecht, Netherlands. 13-50.

Sedgewick R. and Szymanski T. G. (1978). The Complexity of Finding Periods ? *Proceedings of the 11th Symposium on the Theory of Computing.* Atlanta,

Georgia.

Sichel H. S. (1947). Fitting Growth and Frequency Curves by The Method of Frequency Moments. *Journal of the Royal Statistical Society.* **110**, 337-347.

Sichel H. S. (1949). The Method of Frequency Moments and its Application to Type IV Populations. *Biometrika.* **36**, 404-425.

Silverman B. W. (1986). *Density Estimation for Statistics and Data Analysis.* Chapman and Hall. London.

Silver E. A., Pyke D. F. and Peterson R. (1998). *Inventory Management and Production Planning and Scheduling, 3rd Edition.* John Wiley & Sons, New York.

Sohdi M. S. (1999). Developments in ERP, E-Business Rock OR's World. *ORMS TODAY.* **26**, (5) 8.

Stadlober E. (1990). The ratio of uniforms approach for generating discrete random variates. *Journal of Computational and Applied Mathematics*, **31**, 181-189.

Stadlober E. and Zechner H. (1999) The Patchwork Rejection Technique for Sampling from Unimodal Distributions, *ACM Transactions on Modeling and Computer Similation*, **9**, 1, 59-80.

Starr M. K. and Miller D. W. (1962). *Inventory Control: Theory and Practice.* Prentice-Hall, Englewood Cliffs, New Jersey.

Stevenson. W. J. (1982). *Production/Operations Management.* Richard D. Irwin, Inc. Homewood, IL.

Su Honglin (1998). Parameter Analysis on Sedgewicks Cycle Detection Algorithm for Periodic Functions, *http://www.cs.umbc.edu/ hosu/algorithm/tr.html.*

Tashiro Y. (1977). On the methods for generating uniform points on the surface of a sphere. *Annals of Institute of Statistical Mathematics*, **29**, 295-300.

Thannassoulis E., Dyson R. G. and Foster M. J. (1987). Relative efficiency assessments using data envelopment analysis: An application to data on rates departments. *Journal of the Operational Research Society.* Vol. **38**, No. 5, 397-411.

Troutt M. D. (1988). A minimum variance of utility approach to ideal point estimation in certain biased groups. *European Journal of Operational Research*, **35**, 271-277.

Troutt M. D. (1991). A theorem on the density of the density ordinate and an alternative derivation of the Box-Muller method. *Statistics*, **22**, 436-466.

Troutt M. D. (1993). Vertical density representation and a further remark on the Box-Muller method. *Statistics*, **24**, 81-83.

Troutt M. D. (1995). A maximum decisional efficiency estimation principle. *Management Science*, **41**, (1), 77-83.

Troutt M. D., Gribbin D.W., Shanker M. S, and Zhang. A. (2000). Cost Efficiency Benchmarking For Operational Units With Multiple Cost Drivers. *Decision Sciences*, **31**, 4, 813-832.

Troutt M. D., Gribbin D. W., Shanker M. S and Zhang A. (2003a). Maximum Performance Efficiency Approaches For Estimating Best Practice Costs.

Data Mining: Opportunities and Challenges, Edited by J. Wang, 239-259. Idea Group Publishing Co. Hershey, PA., USA.

Troutt M. D., Hu M. and Shanker M. (2001). Unbounded Likelihood in Stochastic Frontier Estimation:Signal-To-Noise Ratio-Based Alternative. *Working Paper*, Department of Management and Information Systems, Kent State University, Ohio.

Troutt M. D., Hu M., Shanker M. and Acar W. (2003b). Frontier versus Ordinary Regression Models for Data Mining. *Managing Data Mining Technologies in Organizations: Techniques and Applications.* Parag C. Pendharkar, ed., Idea Group Publishing Co. Hershey, PA.

Troutt M. D. and Pang W. K. (1997). A further VDR-type density representation based on the Box-Muller method. *Statistics*, **29**, 101-108.

Troutt M. D., Pang W. K. and Hou S. H. (1999). Perfomance of Some Boundary Seeking Mode Estimators on the Dome Bias Model. *European Journal of Operational Research.* **119**, 209-218.

Troutt M. D., Rai A. and Tadisina S. K. (1997a). Aggregating Multiple Expert Data using the Maximum Decisional Efficiency Principle, *Decision Support Systems*, **21**, 75-82.

Troutt M. D., Tadisina S. K. and Sohn, C. (2002). Linear Programming System Identification, working paper, under publication review, Department of Management and Information Systems, Kent State University, Kent, Ohio 44240.

Troutt M. D., Tadisina S. K., Sohn, C. and Brandyberry A. A. (2003c). Linear Programming System Identification, to appear, European Journal of Operational Research.

Troutt M. D., Zhang A., Tadisina S. K. and Rai A. (1997b). Total Factor Efficienty/Productivity Ratio Fitting As An Alternative to Regression and Canonical Correlation Models for Performance Data, *Annals of Operations Research*, **74**, 289-304.

van den Broeck J., Koop G., Osiewalski J. and Steel M. F. J. (1994). Stochastic frontier models: A Bayesian perspective. *Journal of Econometrics.* **61**, 273-303.

Veinott A. F. (1967). The Supporting Hyperplane Method for Unimodal Programming. *Operations Research*, **15**, 147-152.

Verschelde J., Verlinden P. and Cools R. (1994). Homotopies Exploiting Newton Polytopes For Solving Sparse Polynomial Systems. *SIAM J. Num. Anal.* **31**, (3), 915-930.

Vollmann T. E., Berry W. L. and Whybark D. C. (1997). *Manufacturing Planning and Control Systems.* 4th ed., Irwin, Homewood, IL.

Von Neumann J. (1951). *Various Techniques Used in Connection with Random Digits*, NBS Applied Mathematics Series 12, National Bureau of Standards, Washington.

Wakefield J. C., Gelfand A. E. and Smith A. F. M. (1991). Efficient generation of random variates via the ratio-of -uniforms method. *Statistics and Com-*

puting, **1**, 129-133.

Wallace C. S. (1976). Transformed rejection generators for gamma and normal pseudo-random variables. *Australian Computer Journal*, **8**, 103-105.

Walker M. R. (1973). *Determination of the Convex Hull of a Finite Set of Points.* Unpublished M.S. Thesis, Curriculum in Operations Research, University of North Carolina at Chapel Hill.

Weinstock R. (1952). *Calculus of Variations.* McGraw-Hill Book Company, Incorporated, New York.

West M. (1988). Modelling Expert Opinion. *Bayesian Statistics III*, 493-508, J.O. Berger, J.M Bernardo, M.H. DeGoot, and A.F.M. Smith (Eds.), Oxford.

Winkler R. L. (1968). The Consensus of Subjective Probability Distributions. *Management Sci.*, **15** , 61-75.

Winkler R. L. (1986). Comment [on Ganest, C. and Zidek, J. V.], *Statistical Sci.*, **1**, 138-140.

Wolfram S. (1991). *Mathematica: A System for Doing Mathematics by Computer.* 2nd ed. Addison-Wesley, Reading Massachusetts.

Wolfram Research, Inc. (1993). *Mathematica*, Version 2.2, Champaign, Illinois.

Woodard D. L., Chen D. and Madey G. (2002). Cycle Detection in a Chaotic Random Sequence? *Working paper.* Department of Computer Science and Engineering, University of Notre Dame, Notre Dame, Indiana 46556.

Yamaguchi M. and Hata M. (1983). Weierstrass's function and chaos. *Hokkaido Mathematical Journal.* **12**, 333-342.

Yi W. and Bier V. M. (1998). An Application of Copulas to Accident Precursor Analysis. *Management Science.* **44**, (12), Part 2 of 2. S257-S270.

Zangwill W. and Kantor P. B. (1998). Toward a Theory of Continuous Improvement and the Learning Curve. *Management Science.* **44**, (7) July, 910-920.

List of Tables

List of Figures

List of Notations

Selected Notational Conventions

(1) $\mathbf{A}, \mathbf{X}, \mathbf{Y}, \mathbf{A}^{jt}$ are examples of matrices. $\mathbf{A}^{jt}$ denotes a doubly indexed set of matrices.

(2) $\mathbf{A}'$ denotes the transpose of matrix, $\mathbf{A}$. Derivatives of matrices are not used in this book.

(3) $A'(v)$ denotes the derivative of the function, $A(v)$.

(4) $L(\cdot)$ denotes the Lebesgue measure of a set, $(\cdot)$, in $\Re$. $L_d(\cdot)$ denotes the Lebesgue measure of a set, $(\cdot)$, in $\Re^d$, d-dimensional Euclidean space.

(5) ln denotes the natural logarithm function.

(6) x, y, π, η are examples of column vectors.

(7) $X \stackrel{d}{=} Y$ means that random variables, X and Y, have equality in distribution

(8) $\delta_y(x)$ denotes the Dirac delta function with point of support, y.

(9) $\nabla V(\mathbf{x})$ is the gradient vector, the column vector of partial derivatives of the function, $V(x)$.

(10) $I_A(x)$ denotes the indicator function of the set, A. $I_A(x) = 1$ if $x \in A$, and 0 otherwise.

(11) $\Pr(A)$ or $P(A)$ denotes the probability of an event, A.

(12) $\| \mathbf{x} \|$ denotes the Euclidean norm of vector, $\mathbf{x} \in \Re^n$.

(13) cdf or CDF stands for cumulative distribution function

(14) iid stands for independent and identically distributed.

(15) iff stands for if and only if.

(16) $diam(S)$ is the diameter of the set S.

Author Index

Subject Index